T5-AFS-419

DEMCO 38-297

TK
454
M327
1988

132301

N. L. TERTELING LIBRARY
THE COLLEGE OF IDAHO
CALDWELL, IDAHO

Computer-Assisted
Network and
System Analysis

Computer-Assisted Network and System Analysis

E. J. Mastascusa

Bucknell University

WILEY

John Wiley & Sons

New York   Chichester   Brisbane   Toronto   Singapore

TK 454
m 327
1988

Copyright © 1988, by John Wiley & Sons, Inc.

All rights reserved. Published simultaneously in Canada.

Reproduction or translation of any part of this work
beyond that permitted by Sections 107 and 108 of the 1976
United States Copyright Act without the permission of the
copyright owner is unlawful. Requests for permission or
further information should be addressed to the Permissions
Department, John Wiley & Sons.

Library of Congress Cataloging in Publication Data:

Mastascusa, E. J.
  COMPUTER-ASSISTED NETWORK AND SYSTEM ANALYSIS/E. J. Mastascusa.
    p.   cm.
  Bibliography: p.
  Includes index.
  ISBN 0-471-60502-6
  1. Electric circuit analysis—Data processing. I. Title. TK454.M327 1988
621.319'2'0285—dc19   87-29831
                                                                             CIP

Printed in the United States of America

10 9 8 7 6 5 4 3 2 1

132301

Preface

When liberal arts majors begin using computers for word processing and movies are made using computer graphic techniques, we are forced to the realization that computers have indeed invaded every portion of our lives. Paradoxically, some engineers are sometimes slow to integrate computers into their work (but not slow to integrate them into their designs!). Much engineering education fails to integrate computational work as thoroughly as it could be done. Electrical engineering practice is very computerized in such areas as integrated circuit design, but not so much in others. However, programs like SPICE and ACSL are fast becoming ubiquitous. The individual engineer may not write such programs, but he or she is often a heavy user of such programs. Design tools like SPICE and ACSL are so important in many applications that it becomes important for the user to become an educated user.

The minimum goal of this text is to help the reader become an educated user of numerical design tools—particularly circuit analysis programs and simulation programs. Our larger goal is to provide enough basic knowledge that someone can write application-specific programs knowledgeably enough to account for all numerical vagaries and inaccuracies and to produce programs written with good style. Thus, the text is reasonably broad in one sense, covering most of the important topics for electrical engineers, but narrow because many traditional numerical analysis ideas are omitted.

The text assumes that the student is at least at the level of the typical second-semester junior in electrical engineering. Hence the background assumed includes more than just basic circuit analysis. The student is expected to have some familiarity with elementary electronics, linear system analysis (including work with state space concepts), and mathematical work through differential equations. No background in numerical analysis is assumed. The text has been tested with several groups of junior- and senior-level students at Bucknell University.

N. L. TERTELING LIBRARY
THE COLLEGE OF IDAHO
CALDWELL, IDAHO

v

If a text of this kind is to be educationally useful, it is necessary to provide the student with sufficient practice that he or she can become more than just familiar with the concepts. There needs to be serious interaction with detailed examples. The problems and examples in the text assume the possibility of that sort of detailed interaction. To facilitate that interaction, complete programs will be made available, either from the author or from the publisher. The programs that are written at this time include at least the following.

1. A sequence of short programs that exemplify simple concepts, particularly in the first few chapters.
2. A circuit analysis program. This program has five versions, with each version incorporating more concepts from the first three chapters. Because a significant chunk of material is covered in the text, a version of the circuit analysis program incorporates that material, providing more capability. The final version is capable of doing DC and transient analysis of circuits with sources, energy storage elements, transistors, and general controlled sources.
3. An optimization package.
4. A simulation package.
5. A set of routines for generating state equations from topological descriptions of circuits.
6. An interactive control system analysis and design package. This package can compute transient response, set control system parameters, and assist in control system design.

I thank all the students who have offered comments and criticisms concerning the material in the text, particularly those students who were in EE338 through the years. In some cases their names appear in the programs they developed. Some contributed very heavily in the start-up work on the programs. The work of Mark Swartz, Blake Wells, and Kirk Friedman was especially helpful. I was fortunate to have the constant support of my wife, Mary, and my children, Joan, Maria, Diana, Martin, Edward, and Noreen. Many colleagues have lent support or collaborated, particularly Professors R. C. Walker and Maurice Aburdene at Bucknell University and Jim Simes at California State University at Sacramento. To them I offer my gratitude. I also wish to thank several individuals involved in the production of the text, Pete Richardson and Thomas Farrell, the staff of Bucknell Computer Services, and Katharine Rubin of John Wiley. Chris Fyfe, Ruth Miller, and Beverly Spatzer provided assistance in the work and encouragement along the way.

E. J. Mastascusa

Lewisburg, Pennsylvania
October 1987

Contents

Chapter 9 TRANSFER FUNCTION AND CONTROL SYSTEM DESIGN CALCULATIONS 237

Chapter 1

Introduction: Linear Circuits

1.1 INTRODUCTION AND PREVIEW

Computers have gradually invaded many areas of our existence. Programs are available for personal financial calculations, games, writing checks, and balancing checkbooks. In engineering, many special-purpose programs have been written for all kinds of design. Electrical engineers have numerous circuit analysis programs to choose from. They can simulate most of the systems they encounter or design, and they can use those simulations to predict and improve performance.

As design and analysis programs have become more widespread, it has become important for the user to have some knowledge of the limitations and virtues of available programs. There are also still many, many instances in which an individual will have to devise a program to solve some numerical design problem. In either event, some knowledge of the numerical processes involved is necessary. Without that knowledge a real-time calculation may prove impossible to do in real time. Without that knowledge an iterative calculation will not converge, or worse yet, converge to an incorrect answer. Without that knowledge the user may not use his or her time efficiently.

A person who is concerned with numerical solutions of design problems should have some acquaintance with the algorithms that are involved and the range of possible solution methods that exist. In this book we are going to introduce you to many numerical solution methods (algorithms) for many commonly encountered problems in electrical engineering. We will examine ways of calculating what happens in electrical networks, working our way from very simple linear circuits to dynamic nonlinear circuits. We will look at methods for calculating the time behavior of networks, how to optimize a system's performance, and many other algorithms useful to an electrical engineer.

As we proceed through the material, our general approach will be to look at simple examples that illustrate a method, then gradually build on those simple examples to get to the point where realistic problems can be solved. The book assumes that you will gain some experience with numerical solutions as we proceed. FORTRAN programs are provided for all algorithms. (Note, however, that these programs are intended for educational use and are not really a substitute for a well-written program specifically written for the solution of a design problem.) Do as many of the "miniprojects" as possible. They appear at the end of each chapter, and working them through will enable the student to come to grips with the concepts presented in the text.

Computer analysis of circuits and systems is a fascinating and absorbing topic, or at least it can become so. Let us get on with the work.

1.2 GETTING STARTED: DC CIRCUIT ANALYSIS

We begin our exploration of the world of computer-assisted network analysis by considering how we, as humans and as electrical engineers or students, might solve a simple electrical network problem. We may be experienced at solving circuit problems, but we may not be as experienced at solving network problems using computers and computer methods. We will find it to be different than we expect, and different from our human approach to solving this kind of problem.

First, we are going to have to get used to the idea that there is no set, fixed, predetermined way in which we go about producing a computer solution to a problem. For any given problem there are usually either many paths to a solution or none at all. Choosing how to solve a problem digitally is an art, not a science. As we go along, we will try to develop an appreciation of this aspect of the art of computer-aided design.

Computer-aided design and analysis just is not done the same way you would design or analyze circuits or systems. Computer circuit analysis uses different methods than you would use when you do things by hand. Since methods and general approaches are different when you use the computer, you will need to become aware of the differences you have to watch for when you do things on the machine.

As an example, we will consider a simple DC circuit, the one that is shown in Figure 1.1. Here the problem is to find the two node voltages, V_1 and V_2. The way we would go about this as humans is to say:

1. Neglect R_1, R_3, and R_4 since they are small.
2. Assume we then have 1 A (1 ampere) flowing through the 100-Ω (100-ohm) resistor; that is, 100 V/100 Ω.
3. Assume the 1-A current splits between R_3 and R_4, with (1/3) A in R_3, and (2/3) A in R_4.
4. Then we would say

$$V_2 = 0.1 \ \Omega \times (1/3) \ A = 0.0333 \ V$$

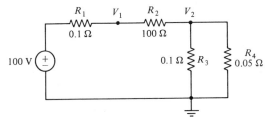

Figure 1.1 An example circuit.

5. And hence

$$V_1 = 100 \text{ V} - (0.1 \ \Omega)(1 \text{ A})$$
$$= 99.9 \text{ V}$$

Now, step back and notice how much insight is required for this approach. Then try to remember the last time you saw a computer exhibit any insight. This solution method is not something that you (or most anyone else) could program on a computer since it depends too much on human insight. Although Hamming [1] said "The purpose of computing is insight, not numbers," when we try to solve a problem using a computer, the computer can only generate numbers; humans still must provide the insight.

To solve the voltage problem, we need to have a method we can program, an algorithm, or set of rules that, if followed, will either generate a solution or indicate that it can't. For the resistor network, a possible method is:

1. Formulate the circuit equations as a set of simultaneous linear equations.
2. Solve the simultaneous linear equations using either a direct method for solution of simultaneous linear equations or a matrix inversion method.

Let us follow this process to see how it works. To formulate the circuit equations in matrix form, we can start by writing Kirchhoff's current law (KCL):

$$\text{Node 1:} \qquad \frac{V_1 - 100}{0.1} + \frac{V_1 - V_2}{100} = 0 \qquad (1.2.1a)$$

$$\text{Node 2:} \qquad \frac{V_2 - V_1}{100} + \frac{V_2}{0.1} + \frac{V_2}{0.05} = 0 \qquad (1.2.1b)$$

These equations may be rearranged to give

$$\text{Node 1:} \qquad V_1(10.01) - V_2(0.01) = 1000 \qquad (1.2.2a)$$
$$\text{Node 2:} \qquad -V_1(0.01) + V_2(30.01) = 0 \qquad (1.2.2b)$$

To use vector matrix form, define V, the voltage vector, and G, the conductance matrix, as follows:

$$V = \begin{bmatrix} V_1 \\ V_2 \end{bmatrix} \qquad (1.2.3)$$

$$G = \begin{bmatrix} 10.01 & -0.01 \\ -0.01 & 30.01 \end{bmatrix} \qquad \cdot (1.2.4)$$

So we would write

$$G * V = S \qquad (* = \text{Matrix multiplication}) \qquad (1.2.5)$$

and the solution for the vector of node voltages is given by

$$V = G^{-1} * S \qquad (1.2.6)$$

One method for solving this set of equations is to take the inverse of the conductance matrix. As humans, we might use the algorithm:

$$G^{-1} = \frac{\text{Adj}\,(G)}{|G|} \qquad (1.2.7)$$

To find the adjoint of G, Adj (G), we form the matrix of cofactors and transpose. For our example:

$$\text{Adj}\,(G) = \begin{bmatrix} 30.01 & 0.01 \\ 0.01 & 10.01 \end{bmatrix}^{\text{T}}$$

$$= \begin{bmatrix} 30.01 & 0.01 \\ 0.01 & 10.01 \end{bmatrix} \qquad (1.2.8)$$

and, $\qquad |G| = (30.01 * 10.01) - (0.01 * 0.01) = 300.4 \qquad (1.2.9)$

So, we end up saying

$$\begin{bmatrix} V_1 \\ V_2 \end{bmatrix} = \begin{bmatrix} 30.01 & 0.01 \\ 0.01 & 10.01 \end{bmatrix} \begin{bmatrix} 1000 \\ 0 \end{bmatrix} * \frac{1}{300.4}$$

$$= \begin{bmatrix} 99.900133 \\ 0.0332889 \end{bmatrix} \qquad (1.2.10)$$

This is fairly close to the answers we got through our approximations. In this case it took longer to get the results manually. For a larger problem, we will be forced to use methods like this in order to have an algorithm that can be programmed. That doesn't mean that we don't have to worry about the algorithm's properties. There may be more than one algorithm that solves any particular problem, just as there's usually more than one way to skin a cat. Then we would have to examine the accuracy and the execution time of possible algorithms. Shortly we will examine some different methods of calculating the matrix inverse and solving simultaneous linear equations as part of the general problem of writing a DC circuit analysis program.

PROBLEM 1.1

For the circuits shown:

(a) Write the node voltage equations and set up the set of linear equations describing the network.

(b) Solve the simultaneous linear equations found in part (a).

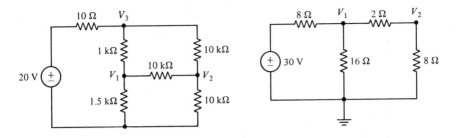

1.3 STARTING TO WRITE A CIRCUIT ANALYSIS PROGRAM

In the preceding section we considered how to find the voltages in a specific, simple circuit. Now, we can begin to define the tasks that must be performed in a circuit analysis program.

What are the tasks that we have to accomplish if we want to write such a program? It looks as if a workable circuit analysis program would have to include at least the following.

1. Input the topological data that describes the circuit interconnections; that is, what element of what value is connected between what nodes?
2. From the topological data, generate a conductance matrix representation for the problem.
3. Solve the conductance matrix equations.

Each of these tasks is separate from the others and will normally be a separate part of a program. When we program each task, we will do it separately; that is, we will modularize our program so that each separate task is performed by a separate subroutine. We should also note that there are several different approaches that can be used to construct such a program. We could build up a description of a circuit in a computer file (Task 1) and then later run a program that will access that data. Alternatively, the program could be interactive, permitting the user to describe and alter circuit data, making decisions as computations are done.

The popular circuit analysis program, SPICE, illustrates how topological information is important. A SPICE source program is shown in Table 1.1. Notice how the node connections for each element are an integral part of the description of each element.

In Appendix A you will find a simple DC circuit analysis program. That program is interactive. The user can enter a circuit, modify or eliminate elements, and then solve for node voltages. The construction of such a program involves keeping a copy of the present circuit data in a set of arrays. For that program the arrays that are used are:

NODES(I, J) contains information on node connections of elements.

VALUES(I) contains the value of element I in appropriate units.

TABLE 1.1   A SPICE EXAMPLE

```
 SPICE EXAMPLE 1
R1  1  2  1000
R2  2  3  1000
C1  2  0  1UF
C2  3  0  1UF
VIN  1  0  PULSE(0  10)
•TRAN 100USEC 5MSEC
•PRINT TRAN V(1) V(2) V(3)
•END
```

KINDS(I) contains the kind of element I in literal code.

SUPP (I) contains supplementary information for an element (initial conditions for L and C, $R_{internal}$ for E).

To enter information into these arrays interactively, a program segment of the sort shown in Figure 1.2 is used. This approach allows the user to build a circuit description interactively. After requesting a literal input, the input is checked to determine what type of element is being added, then prompts for resistance, current, or voltage values.

No matter what particular program we use, it seems apparent that this kind of topological description is where we must begin. That is probably equivalent to a human having a circuit diagram to work with. However, with the topological information safely stored in these arrays, we still do not have the information in a form that we can use to get a solution. Our first goal has to be

```
      IF (REPLY .EQ. 'R') THEN
          NELMT = NELMT + 1
          WRITE (6,*)' NODES          RESISTANCE VALUE (KOHMS)'
          READ  (5,*) NODES(NELMT,1), NODES(NELMT,2), VALUES(NELMT)
          VALUES(NELMT) = VALUES(NELMT)*1000.
          SUPP(NELMT) = 0.
          KINDS(NELMT) = REPLY
      ENDIF
      IF (REPLY .EQ. 'R') GO TO 1
C
      IF (REPLY .EQ. 'E') THEN
          NELMT = NELMT + 1
          WRITE (6,*)' NODES     VOLTAGE     INTERNAL RESISTANCE (KOHMS)'
        READ(5,*)NODES(NELMT,1),NODES(NELMT,2),VALUES(NELMT),SUPP(NELMT)
          SUPP(NELMT) = SUPP(NELMT)*1000.
          KINDS(NELMT) = REPLY
      ENDIF
      IF (REPLY .EQ. 'E') GO TO 1
C
      IF (REPLY .EQ. 'I') THEN
          NELMT = NELMT + 1
          WRITE (6,*)' NODES        CURRENT'
          READ  (5,*) NODES(NELMT,1), NODES(NELMT,2), VALUES(NELMT)
          KINDS(NELMT) = REPLY
      ENDIF
      IF (REPLY .EQ. 'I') GO TO 1
```

Figure 1.2 Data input program segment.

to get the data into a conductance matrix form so that we have all of the numbers in the vector matrix equation that represents this circuit; that is,

$$G * V = S \qquad (1.2.5)$$

Let's look back to the set of equations we wrote in Section 1.1. There we wrote Kirchhoff's current law for each of the two nodes that had unknown node voltages. In other words, we set up equations for the unknown voltages, which were V_1 and V_2. We had no need to write an equation for the node, which we knew was at 100 V.

Now, if we neglect the fact that we knew the voltage at one node, we can note that when we write Kirchhoff's current law at any node, the coefficient of the voltage at that node is the sum of the conductances (reciprocal resistances) attached to that node. Second, the coefficients of the other voltages are the negative sums of all the conductances attached to both nodes (shared between two nodes).

Figure 1.3 gives a program fragment that will generate a conductance matrix from topological information in the form we are using. This program fragment takes the information contained in NODES, VALUES, and KINDS (all are arrays) and generates a conductance matrix G (a square array) by determining which resistances (element type R) are connected to a particular node (I or J). Note the following:

1. We assume that the conductance matrix G is filled with zeros at the start of the program.
2. We assume that the ground node is labeled 0. (This convention is used in other programs—SPICE, for example.)

Voltage sources can be a problem in this kind of formulation. Every grounded source fixes a node at some definite voltage with respect to ground. An ungrounded source fixes a voltage difference between two nodes. This is a problem because we have to keep track of which node voltages are known, and which are unknown, as well as which node voltage differences are set. If we write a Kirchhoff's current law equation at each node, we will have more equations than unknowns.

```
      DO 100 I=1,NELEMENTS
       IF (KINDS(I).EQ. 'R' ) THEN
        IF(NODES(I,1).NE.0) THEN
         G(NODES(I,1),NODES(I,1))=G(NODES(I,1),NODES(I,1))+1./VALUES(I)
         IF(NODES(I,2).NE.0) THEN
         G(NODES(I,2),NODES(I,2))=G(NODES(I,2),NODES(I,2))+1./VALUES(I)
         G(NODES(I,1),NODES(I,2))=G(NODES(I,1),NODES(I,2))-1./VALUES(I)
         G(NODES(I,2),NODES(I,1))=G(NODES(I,2),NODES(I,1))-1./VALUES(I)
         ENDIF
        ELSE
         IF(NODES(I,2).NE.0) THEN
         G(NODES(I,2),NODES(I,2))=G(NODES(I,2),NODES(I,2))+1./VALUES(I)
         ENDIF
        ENDIF
       ENDIF
100 CONTINUE
```

Figure 1.3 Conductance matrix program segment.

One solution to this problem is to require all voltage sources to have a nonzero internal resistance. An internal resistance establishes a node inside the resistor/voltage-source combination. That node can be eliminated by doing a conversion from voltage source to current source (Thévenin to Norton). We would be left with only nodes with unknown voltages—a configuration with as many node equations as unknown voltages. We will adopt the convention that all voltage sources have internal resistances. Then we can do a Thévenin-to-Norton conversion and add the proper currents to all nodes to which that source is attached.

The program fragment in Figure 1.2 was designed to prompt for an internal resistance for a voltage source. Forcing each voltage source to be accompanied by an internal resistance then permits easy numerical formulation of a conductance matrix and source vector using a subroutine such as the one shown in Figure 1.4.

TOPOTOCOND takes the topological information in the arrays NODES, VALUES, KINDS, and SUPP, and fills in a conductance matrix G and a source vector SOURCES.

```
      SUBROUTINE TOPOTOCOND(NELEMENTS,G,SOURCE)
C
      COMMON/CIRCUIT/NODES(20,4),VALUES(20),SUPP(20)/CHAR/KINDS(20)
      DIMENSION G(20,20), SOURCE(20)
      CHARACTER*1 KINDS
C
C ZERO CONDUCTANCE MATRIX AND SOURCE VECTOR
C
      DO 10 I=1,10
      SOURCE(I) = 0.
      DO 5  J = 1,10
      G(I,J) = 0.
   5  CONTINUE
  10  CONTINUE
C
C THIS SUBROUTINE GENERATES A CONDUCTANCE MATRIX AND SOURCE VECTOR
C FROM TOPOLOGICAL INFORMATION CONTAINED IN NODES, VALUES AND KINDS.
C
      DO 100 I=1,NELEMENTS
       IF (KINDS(I).EQ. 'R' ) THEN
        IF(NODES(I,1).NE.0) THEN
        G(NODES(I,1),NODES(I,1))=G(NODES(I,1),NODES(I,1))+1./VALUES(I)
        IF(NODES(I,2).NE.0) THEN
         G(NODES(I,2),NODES(I,2))=G(NODES(I,2),NODES(I,2))+1./VALUES(I)
         G(NODES(I,1),NODES(I,2))=G(NODES(I,1),NODES(I,2))-1./VALUES(I)
         G(NODES(I,2),NODES(I,1))=G(NODES(I,2),NODES(I,1))-1./VALUES(I)
        ENDIF
       ELSE
        IF(NODES(I,2).NE.0) THEN
         G(NODES(I,2),NODES(I,2))=G(NODES(I,2),NODES(I,2))+1./VALUES(I)
        ENDIF
       ENDIF
      ENDIF
 100 CONTINUE
C
C NOW GENERATE THE SOURCE VECTOR
C
C
```

Figure 1.4 Subroutine to generate a conductance matrix from topological information.

```
      DO 200 I = 1,NELEMENTS
C
C CHECK TO SEE IF WE HAVE FOUND A VOLTAGE SOURCE
C
        IF(KINDS(I).EQ. 'E' ) THEN
        IF(NODES(I,1).NE.0) THEN
         G(NODES(I,1),NODES(I,1))=G(NODES(I,1),NODES(I,1))+1./SUPP(I)
         SOURCE(NODES(I,1)) = SOURCE(NODES(I,1)) + VALUES(I)/SUPP(I)
         IF(NODES(I,2).NE.0) THEN
          G(NODES(I,2),NODES(I,2))=G(NODES(I,2),NODES(I,2))+1./SUPP(I)
          G(NODES(I,1),NODES(I,2))=G(NODES(I,1),NODES(I,2))-1./SUPP(I)
          G(NODES(I,2),NODES(I,1))=G(NODES(I,2),NODES(I,1))-1./SUPP(I)
          SOURCE(NODES(I,2)) = SOURCE(NODES(I,2)) - VALUES(I)/SUPP(I)
         ENDIF
        ELSE
         IF(NODES(I,2).NE.0) THEN
          G(NODES(I,2),NODES(I,2))=G(NODES(I,2),NODES(I,2))+1./SUPP(I)
          SOURCE(NODES(I,2)) = SOURCE(NODES(I,2)) - VALUES(I)/SUPP(I)
         ENDIF
        ENDIF
        ENDIF
  200 CONTINUE
C
C NOW SEARCH FOR ALL CURRENT SOURCES
C
      DO 300 I = 1,NELEMENTS
        IF(KINDS(I).EQ. 'I' ) THEN
         IF(NODES(I,1).NE.0) THEN
           SOURCE(NODES(I,1))=SOURCE(NODES(I,1))+VALUES(I)
         ENDIF
         IF(NODES(I,2).NE.0) THEN
           SOURCE(NODES(I,2))=SOURCE(NODES(I,2))-VALUES(I)
         ENDIF
        ENDIF
  300 CONTINUE
C
      RETURN
      END
```

Figure 1.4 (*continued*)

The steps in our algorithm to generate G and SOURCES are summarized as follows:

1. We zero all the elements of G and SOURCES.
2. If a resistor R is connected between two nodes, A and B, (see Figure 1.5) then we increment $G(A, A)$ and $G(B, B)$ by $1/R$ and decrement $G(A, B)$ and $G(B, A)$ by $1/R$.
3. If a resistor R is connected between node A and ground (and the ground node is *required* to be node 0), we increment $G(A, A)$ by $1/R$. Nothing else is done.
4. If a voltage source is found, we convert it to a current source in parallel with the internal resistance. Then we use rules 2 and 3 above to change the conductance matrix G.
5. The value of the current in the current source is added to the node into

Figure 1.5 A resistor.

which the current flows and subtracted from the node out of which it flows. If one node is the ground node, no operation is performed.

With these programs in hand, we have recast our problem into one in which we have to solve $G * V = $ SOURCES (or just $G * V = S$). The solution of this set of equations ($G * V = $ SOURCES) can be thought of in at least two different ways. We could explicitly take the inverse of the conductance matrix G using the solution $V = G^{-1} * S$. We could also view the problem as a simultaneous-equation problem instead of a matrix inversion problem and attempt to find a simpler solution. In what follows we will first examine matrix inversion methods (which are the most general methods) and later develop methods specifically designed to solve just simultaneous linear equations. These two problems are very closely related, although they differ somewhat in execution time. Both methods will be needed at various points later, so we consider them both at this point.

1.4 NUMERICAL SOLUTION OF SIMULTANEOUS LINEAR EQUATIONS

In our DC circuit analysis program, we must ultimately solve the vector matrix equation $G * V = S$ for the vector of node voltages, V. One way to do so is to compute the inverse of the conductance matrix G^{-1} and then multiply G^{-1} by the source vector S. Alternatively, we can consider the problem to be the solution of the simultaneous linear equations $G * V = S$. It will usually be the case that less CPU time is needed if only the linear-equation solution is needed, and no explicit calculation of the matrix inverse has to be done. We can consider both cases because there are times when a matrix inverse is needed and other times when linear equation solutions will suffice.

For the moment, imagine that we wanted to compute the inverse of the conductance matrix. We would need to estimate how much computation that involves and how the amount of computation depends upon the number of simultaneous equations. Consider computing the inverse of a square matrix, M. Formally, we might take:

$$M^{-1} = \frac{\text{Adj}(M)}{|M|} \qquad (1.4.1)$$

where

$$M^{-1} = \text{Inverse of } M \qquad (1.4.2a)$$
$$\text{Adj}(M) = \text{Adjoint of } M$$
$$= \text{Transpose of matrix of cofactors} \qquad (1.4.2b)$$
$$|M| = \text{Determinant of } M \qquad (1.4.2c)$$

We could, conceivably, use this algorithm to calculate an inverse. There are some dangers here, however. In order to calculate the adjoint, we would have to calculate N^2 cofactors (since the adjoint is the transpose of the matrix of cofactors). We can estimate the amount of computation necessary for calculation of each cofactor since a cofactor is a determinant, of size $(N - 1)$ by $(N - 1)$—when the matrix is $N \times N$. For a 2 by 2 determinant, we have two multiplications (and a subtraction) to do to evaluate its determinant. Going to the next size, if we expanded a 3 by 3 determinant by cofactors, we would have to do three multiplications of cofactors, and each cofactor would have taken two multiplications, for a total of 6. In general, evaluation of an $(N - 1)$ by $(N - 1)$ determinant with this algorithm will take $(N - 1)!$ multiplications. That can be a large number! The factorial function grows faster than just about any other function—an undesirable property if we want to solve large problems.

PROBLEM 1.2

Assume you can do a floating-point multiplication in 1 μs (1 microsecond) on a computer. How much time would it take to compute a determinant of a 10×10 matrix? A 15×15? A 20×20?

We need to consider something other than this "human-oriented" algorithm. Many times the algorithms used to illustrate and drill a concept are not really useful for more complex, realistic problems.

Let us consider one of the popular methods of solving simultaneous linear equations, the Gaussian elimination method. To illustrate this algorithm we will choose a simple example, and follow it through.

EXAMPLE 1.1

Consider solving the two simultaneous linear equations. The example we will use is the set of two simultaneous equations:

$$10x + \quad y = 17$$
$$2x + 20y = 23$$

Multiply the first line by 0.1:

$$x + 0.1y = 1.7$$
$$2x + \quad 20y = 23$$

Subtract twice the first line from the second line:

$$x + 0.1y = 1.7$$
$$19.8y = 23 - 3.4$$

Multiply the second line by (1/19.8):

$$x + 0.1y = 1.7$$
$$y = \frac{19.6}{19.8}$$

Subtract (0.1) times the second line from the first line:

$$x \qquad = 1.7 - 0.1\left(\frac{19.6}{19.8}\right)$$

$$y = \frac{19.6}{19.8}$$

This approach seems straightforward and is easily programmed. A program implementing this basic algorithm is given in Figure 1.6. (The subroutine is called LINGAU for LINear equation solution using GAUssian elimination.) Any practical program, however, must recognize certain kinds of difficulties that can arise. Let us look at one of the problems.

Consider computing the solution of three simultaneous equations. The next example is a set of equations that gives no problems.

EXAMPLE 1.2

Consider the equations

$$\begin{aligned}
2x + y + z &= 1 \\
x + 2y + z &= 2 \\
x + y + 2z &= 3
\end{aligned}$$

Use Gaussian elimination to solve this set of equations.

1. Multiply the first equation by 0.5 and use the result to eliminate x in the other two equations. We get

$$\begin{aligned}
x + 0.5y + 0.5z &= 0.5 \\
1.5y + 0.5z &= 1.5 \\
0.5y + 1.5z &= 2.5
\end{aligned}$$

2. Multiply the second equation by $\frac{2}{3}$ and use the result to eliminate y in the last equation. We get

$$\begin{aligned}
x + 0.5y + .5z &= 0.5 \\
y + (1/3)z &= 1.0 \\
(4/3)z &= 2.0
\end{aligned}$$

3. Solve for z. Substitute back to get y and x.

$$\begin{aligned}
x &= -.5 \\
y &= .5 \\
z &= 1.5
\end{aligned}$$

A very similarly structured example that presents a problem in solution is the next one.

```
C
C***************************************************************
C
      SUBROUTINE LINGAU (A,RHS,N)
C
C LINGAU USES GAUSSIAN REDUCTION TO SOLVE A SET OF LINEAR
C SIMULTANEOUS EQUATIONS, A*X = RHS.
C
C Version = July 24,1985
C
      DIMENSION A(20,20), RHS(20)
C
C CHECK TO SEE IF A IS JUST 1 BY 1.
C
            IF (N .EQ. 1) THEN
                RHS(1) = RHS(1)/A(1,1)
                RETURN
            ENDIF
      DO 100 M = 2,N
      M1 = M-1
         IF(A(M1,M1) .EQ. 0.) THEN
                CALL MATPIVOT (A,RHS,M1,M1,N,N)
         ENDIF
C
C ZERO OUT COLUMN ELEMENTS BELOW DIAGONAL
C
      AM = 1./A(M1,M1)
      DO 100 K = M,N
      AK = -A(K,M1)*AM
         DO 20 L = M1,N
   20    A(K,L) = A(K,L) + AK*A(M1,L)
C
      RHS(K) = RHS(K) + AK*RHS(M1)
C
  100 CONTINUE
C
      N1 = N-1
      DO 300 M = 1,N1
      J = N-M+1
      AM = 1./A(J,J)
      M1 = J-1
         DO 290 K = 1,M1
         J1 = J-K
         AK = -AM*A(J1,J)
         A(J1,J) = 0.
            RHS(J1)  = RHS(J1)  + AK*RHS(J)
  290    CONTINUE
  300 CONTINUE
C
      DO 400 M = 1,N
      AM = 1./A(M,M)
  390    RHS(M)  = RHS(M) *AM
  400 CONTINUE
      RETURN
      END
```

Figure 1.6 Subroutine for simultaneous linear equation solution using Gaussian reduction.

EXAMPLE 1.3

Consider solving the equations

$$\begin{aligned}
2x + y \quad &= 1 \\
2x + y + z &= 2 \\
x + y + z &= 3
\end{aligned}$$

Now, divide the first equation by 2 and use the result to eliminate x in the other equations.

$$x + y/2 \;\; = 0.5$$
$$z = 1.0$$
$$y/2 + z = 2.5$$

There is a problem here. The y term in the second equation is already eliminated. If we interchange the second and third equation, we will have

$$x + y/2 \;\;\;\; = 0.5$$
$$y/2 + z = 2.5$$
$$z = 1.0$$

If we choose to interchange these two equations, we will be able to continue with a solution.

Clearly when a zero arises on the diagonal, rows should be switched, if possible. This process is called pivoting, and is implemented in the subroutine in Appendix A. For accuracy reasons, the row of choice to interchange with the zero-diagonal row is the row with the largest element.

Using Gaussian elimination, the solution of simultaneous linear equations involves simple operations like multiplication of rows by constants, and addition and subtraction of rows. The same approach can be used to calculate a matrix inverse. Imagine that we have a matrix, M, and we want to compute its inverse. We will set up another matrix, MI, and originally MI will be an identity matrix. Then we will perform operations on both M and MI and the result will be to have MI contain the inverse of M. All of the operations we will do will be the ones used in Gaussian elimination.

We will go through the sequence of operations just used to solve the set of equations, and whatever we do to M we will also do to MI. Now, just follow along and see what happens as we do that. (Note, we will use "subscripts" on M and MI to denote different versions.)

$$\begin{bmatrix} 10 & 1 \\ 2 & 20 \end{bmatrix} = M_0 \qquad \begin{bmatrix} 1 & 0 \\ 0 & 1 \end{bmatrix} = MI_0 \qquad (1.4.3)$$

We will operate on M, and simultaneously operate on MI, doing the same to MI as is done to M.

1. Divide the top row by 10:

$$\begin{bmatrix} 1 & 0.1 \\ 2 & 20 \end{bmatrix} = M_1 \qquad \begin{bmatrix} 0.1 & 0 \\ 0 & 1 \end{bmatrix} = MI_1 \qquad (1.4.4a)$$

2. Subtract twice the top row from the bottom row:

$$\begin{bmatrix} 1 & 0.1 \\ 0 & 19.8 \end{bmatrix} = M_2 \qquad \begin{bmatrix} 0.1 & 0 \\ -0.2 & 1 \end{bmatrix} = MI_2 \qquad (1.4.4b)$$

3. Divide the second row by 19.8:

$$\begin{bmatrix} 1 & 0.1 \\ 0 & 1 \end{bmatrix} = M_3 \qquad \begin{bmatrix} 0.1 & 0 \\ -0.2/19.8 & 1/19.8 \end{bmatrix} = MI_3 \qquad (1.4.4c)$$

4. Subtract one-tenth of the second row from the first row:

$$\begin{bmatrix} 1 & 0 \\ 0 & 1 \end{bmatrix} = M_4 \qquad \begin{bmatrix} 2 & -0.1 \\ -0.2 & 1 \end{bmatrix} * \frac{1}{19.8} = MI_4 \qquad (1.4.4d)$$

If we look carefully at MI_4, we will see that it is the inverse of M. (It may take some careful looking and calculation, but it is true. Multiply MI_4 by M to check whether MI_4 is indeed the inverse of M.)

Why is this the inverse? Look at what we do in the various steps. Summarize the steps:

1. Divide the top row by 10.
2. Subtract twice the top row from the bottom row.
3. Divide the second row by 19.8.
4. Subtract one-tenth of the second row from the first row.

The first step can be thought of as a matrix multiplication operation; that is,

$$\begin{bmatrix} 0.1 & 0 \\ 0 & 1 \end{bmatrix} \times \begin{bmatrix} 10 & 1 \\ 2 & 10 \end{bmatrix} = \begin{bmatrix} 1 & 0.1 \\ 2 & 20 \end{bmatrix} = M_1 \qquad (1.4.5)$$

Denote the matrix premultiplying M_1 by P_1. Then we have

$$M_2 = P_1 \times M_1 \qquad (1.4.6)$$

All four of the steps above work out to be expressible as a matrix multiplication, with

$$M_4 = P_4 \times P_3 \times P_2 \times P_1 \times M_1 \qquad (1.4.7)$$

where

$$P_1 = \begin{bmatrix} 0.1 & 0 \\ 0 & 1 \end{bmatrix} \qquad P_2 = \begin{bmatrix} 1 & 0 \\ -2 & 1 \end{bmatrix}$$

$$P_3 = \begin{bmatrix} 1 & 0 \\ 0 & 1/19.8 \end{bmatrix} \qquad P_4 = \begin{bmatrix} 1 & -0.1 \\ 0 & 1 \end{bmatrix} \qquad (1.4.8)$$

But since

$$M_0^{-1} \times M_0 = I \qquad (1.4.9)$$

then we must have

$$M_0^{-1} = P_4 \times P_3 \times P_2 \times P_1 \times I \qquad (1.4.10)$$

In other words, if we reduce a matrix M to an identity matrix using Gaussian reduction, and we do the same set of operations on an identity matrix, we can generate the inverse.

One question that arises is how much computer time this will take. One way of comparing different algorithms is to estimate the number of operations (multiplications, divisions, additions, subtractions) the algorithm does when it executes. Assume we have an $N \times N$ system to solve. For the Gaussian reduction method, we have

1. N divisions to get a 1 in the first row.
2. N multiplications and subtractions to get the zero in the leading column of each row, or $N \times (N-1)$ multiplications and subtractions.

So, we have a total of $(N \times (N-1)) + N$, or N^2, operations, just to do the first column. If we want to zero all the elements below the diagonal, and put ones on the diagonal, we will need:

$$N^2 + (N-1)^2 + \cdots (2)^2 + (1)^2 \text{ multiplications} \qquad (1.4.11)$$

This sum has the value:

$$\frac{N(N+1)(2N+1)}{6} \qquad (1.4.12)$$

To zero the upper part of the matrix will require an equal number of operations, so the total is

$$\frac{N(N+1)(2N+1)}{3} \qquad (1.4.13)$$

These two options correspond to interchanging columns and rows in our matrices, M and MI. Again, as long as we do the same thing to both M and MI, our algorithm will get to the inverse, as long as it exists. There are a few pitfalls here. We might have a zero on the diagonal that would prevent the division that needs to be done to get a one on the diagonal. If we have that situation, we can interchange two rows, or two columns. However, we must be cognizant of this possibility, and be prepared to pivot by rows or columns, as necessary.

Our programs include the possibility of pivoting. Figure 1.7 shows those programs for Gaussian inversion with pivoting. There is some strategy involved in choosing when and how to pivot. The simplest strategy would be to pivot only when absolutely necessary. That would be when we encounter a zero element on the diagonal, and we are unable to use that element to zero out the rows below. That would occur whenever we have the situation

```
1  X  X  X  X  X  X
0  1  X  X  X  X  X
0  0  1  X  X  X  X
0  0  0  0  X  X  X
0  0  0  X  X  X  X
0  0  0  X  X  X  X
0  0  0  X  X  X  X
```

```
C
C*****************************************************
C
C
      SUBROUTINE MATPIVOT (A,RHS,I,J,N,M)
C
C MATPIVOT SELECTS THE MAXIMUM ELEMENT
C FROM I TO N, AND J TO M
C AND PIVOTS THAT ELEMENT TO POSITION (I,J).
C
C Version = July 24, 1985
C
      DIMENSION A(20,20),RHS(20)
C
      AMAX = ABS(A(I,J))
      IMAX = I
      JMAX = J
C
      DO 10 MI=I,N
      DO 10 MJ=J,M
C
      IF (ABS(A(MI,MJ)) .GT. AMAX) THEN
            AMAX = ABS(A(MI,MJ))
            IMAX = MI
            JMAX = MJ
      ENDIF
   10 CONTINUE
C
C EXCHANGE ROWS IF NECESSARY
C
      IF (IMAX .GT. I) THEN
            DO 20 MI=1,N
               ATEMP = A(I,MI)
               A(I,MI) = A(IMAX,MI)
               A(IMAX,MI) = ATEMP
   20       CONTINUE
            RTEMP = RHS(I)
            RHS(I) = RHS(IMAX)
            RHS(IMAX) = RTEMP
      ENDIF
C
      RETURN
      END
```

Figure 1.7 Subroutine for Gaussian inversion with pivoting.

Here the zero-diagonal element would stop us unless we can pivot. In this case we have to pivot. However, we might encounter a small, but nonzero, diagonal element. In that case it might be advantageous to pivot. Actually, it is always worthwhile pivoting to bring the largest possible element up to the diagonal [2].

ALGORITHM

GAUSSIAN REDUCTION SOLUTION OF LINEAR EQUATIONS

The Gaussian Reduction method can be used to solve a set of N simultaneous linear equations for N unknowns. A brief description of the algorithm is as follows:

1. The N equations should be arranged with all variables in the same order in each equation (as in all of the examples). Normalize the first row by dividing all coefficients by the coefficient of the first variable. Divide the right-hand side as well.

$$a'_{i,j} \leftarrow a_{i,j}/a_{1,1} \qquad j = 1, \ldots, N \qquad b'_1 \leftarrow b_1/a_{1,1}$$

2. Zero out the remainder of the first column.

$$a'_{i,j} \leftarrow a_{i,j} - a_{i,1}a_{1,j} \begin{cases} i = 2, \ldots, N \\ j = 1, \ldots, N \end{cases} \qquad b'_i \leftarrow b_i - b_1 a_{i1}$$

3. If the coefficient of the first variable is zero, then
   (a) Interchange rows if a coefficient in the first column is nonzero.

   $$a'_{i_1,j} \leftarrow a_{i_2,j}, \qquad a'_{i_1,j} \leftarrow a_{i_2,j} \qquad \begin{cases} a_{i_2,j} \neq 0 & b'_{i_1} \leftarrow b_{i_2} \\ i = 1 - N & b'_{i_2} \leftarrow b_{i_1} \end{cases}$$

   (b) Interchange columns to pivot a nonzero element into the $1 - 1$ position.

   $$a'_{i,j_1} \leftarrow a_{i,j_2}, \qquad a'_{i,j_2} \leftarrow a_{i,j_1} \begin{cases} a_{i,j_2} \neq 0 \\ i = 1 - N \end{cases}$$

   (c) If no nonzero element can be pivoted to the position of the first coefficient, then no solution exists and an error message should be given to the user.
4. Continue to the next diagonal coefficient, treating it as the first coefficient, redoing steps 1, 2, and 3 until no more diagonal elements remain.

1.5 AC CIRCUIT ANALYSIS

Sinusoidal signals are the most common form of excitation in many circuit applications. Sinusoidal signals are found in power systems and in communication systems. They are also commonly used as test signals in audio equipment and control systems. Sinusoidal signals have unique properties that make them useful in linear systems. If a sinusoidal signal drives a system, the response of the system will also be sinusoidal. Sinusoidal signals, in that sense, are the eigenfunctions for linear systems. They retain their form when acted upon by linear systems, changing only amplitude and phase, retaining frequency and the essential sinusoidal character.

The methods developed in this chapter can easily be extended to cover linear circuits with sinusoidal excitation—that is, sinusoidal voltage and current sources. Here we will consider some details of what needs to be done to extend those techniques.

Let us momentarily review the steps needed to do computer analysis of linear circuits as they were developed in this chapter. Those tasks are as follows:

1. The program must store a topological description of the network, including element type, interconnection data, and numerical data for each element.

2. From the network description, a conductance matrix representation must be generated.

3. Node voltages must be solved for from the conductance matrix representation.

We can consider how each step has to be modified in order to accommodate the "problem elements" in AC analysis—that is, capacitors and inductors.

There is no particular problem storing the topological data for capacitors and inductors. In fact, in Chapter 3 that same problem is encountered when we consider computation of the transient response of circuits with capacitive and inductive elements. All that is necessary is to establish an L and a C in the set of codes for element types, and to be able to recognize those types as necessary within the program.

The significant change that occurs when capacitors and inductors are added is that complex analysis must be used to handle the algebra that is done to compute sinusoidal steady-state solutions for circuits with AC excitation. The concepts of impedance and admittance (instead of resistance and conductance) are necessary in AC analysis, and the reader needs to be familiar with those concepts (which are usually introduced in a first course on linear circuit and system analysis). For these elements we have

$$Z_c = 1/j\omega C \qquad Y_c = j\omega C$$
$$Z_l = j\omega L \qquad Y_l = 1/j\omega L$$

The complicating issue here is that these impedances and admittances are all imaginary, and the algebra that results in circuit analysis is all complex. Thus, instead of dealing with real conductance matrices, we must now deal with complex admittance matrices. In addition, the voltages and currents are "phasors"— that is, complex quantities that code the amplitude and phase information about each voltage and current in the circuit.

The set of equations that results from AC analysis will be of the form:

$$Y * V = I$$

where Y is the complex admittance matrix. Solving a set of simultaneous, complex linear equations is, in principle, no different from solving a set of simultaneous real linear equations. Programming the solution can be another matter. Ralston [3] has a few pertinent comments on the subject:

> The most efficient way to solve the system of linear equations with complex elements is to follow the real algorithm, replacing all real operations by complex ones. This can readily be accomplished in a programming language which allows for the declaration of complex variables. In the absence of such a facility, it can become tiresome to convert every real operation into a series of such operations.

Fortunately, FORTRAN readily permits declaration of complex variables and grants access to a large library of complex algebraic functions.

1.6 WHERE PROBLEMS CAN ARISE

There are clearly going to be situations in which problems can arise when trying to get a numerical solution to a set of simultaneous linear equations. To see the kind of difficulty, consider a case in which no solution exists.

$$x + y = 1$$
$$x + y = 2$$

These equations can be interpreted geometrically. Each equation is the equation of a straight line. For the two simultaneous equations above, the two straight lines are parallel, the determinant of the system is zero, and no solution exists.

Now, let us consider what happens if we change things just slightly. Consider the set of equations

$$x + \quad y = 1$$
$$x + 1.1y = 2$$

This set of equations has a solution. Again, the situation can be visualized geometrically. The two straight lines defined by our new simultaneous equations are shown in Figure 1.8 at the right. The original parallel lines are shown at the left. In the second case, the two lines are not parallel, but they meet at a small angle. What happens if we change things so that the angle becomes even smaller? Change the coefficient of 1.1 to a value of 1.05. Figure 1.9 shows what happens then.

That very small change in the coefficient changes the solution dramatically. A further change from 1.05 to 1.0 will cause the solution to cease to exist entirely. The solution of this system of equations is very sensitive to changes in the coefficients, and is said to be "ill conditioned."

One might naively expect that it would be difficult to devise a circuit that exhibits this kind of numerical sensitivity to parameter variations. However, it is not difficult at all. In Figure 1.10 we have a circuit in which two 2-kΩ resistors are connected by 1-Ω resistor. We can set up the Kirchhoff current law equations for the network.

$$(0.001 + 0.0005 + 1)V_1 - V_2 = (0.001)10$$
$$-V_1 + (0.0005 + 1)V_2 = 0$$

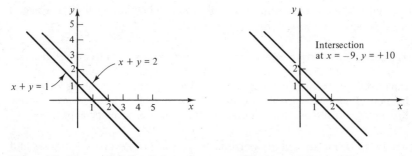

Figure 1.8 The graphical solution of two sets of linear equations.

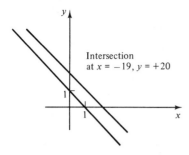

Figure 1.9 A case of nearly parallel lines.

The determinant of this set of equations becomes:

$$(1.0015 * 1.0005) - 1.00000000 = 1.00200075 - 1. = 0.00200075$$

This is the difference between two nearly equal numbers, and a slight amount of roundoff error can seriously affect the result. We should also note that the problem will become considerably worse if the 1-Ω resistor becomes smaller or if any of the other resistors becomes significantly larger.

Consider what happens if the 1-Ω resistor stays fixed and all of the other resistors are made larger. As that happens, the coefficient of V_1 in the KCL equation at node 1 will approach unity. In the limit the KCL equations approach the following form.

$$V_1 - V_2 = \frac{10 \text{ V}}{\text{large resistance}}$$
$$-V_1 + V_2 = 0 \text{ V}$$

This set of equations is indeterminate, and has no solution. The real problem is that sooner or later, as those resistances are increased, the coefficients of V_1 (in the first KCL equation) and of V_2 (in the second KCL equation) become so close to unity that the machine representation of those numbers is unity because of the finite length of the number representation in any digital computer. At that point the problem becomes unsolvable. On the way to that point, solutions become increasingly unreliable.

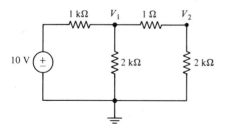

Figure 1.10 A network with an ill-conditioned conductance matrix.

PROBLEM 1.3

In many cases, the reason a small resistor will enter a circuit is that the small resistor is the internal resistance in a voltage source. The circuit shown is an example. Determine whether the solution for voltages in this network is sensitive to the value of the internal resistance of the source in the way that the preceding network was sensitive to the value of the small resistor.

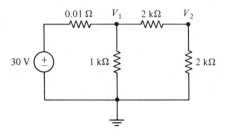

It seems as though the problem centers on those situations where there is a small resistance between two nodes. In that case, in one sense, the two nodes are approaching being a single node. If the resistor is tending to be small or close to zero, we will have problems. As the resistance approaches zero, we find a point at which we begin seeing problems. When the resistance goes to zero, and the nodes coalesce, the numerical problems may well disappear. However, in between there is a never-never land of which the informed analyst will be wary.

The problem of inaccuracy in solution of linear systems of equations has received a great deal of attention over the years. Chapter 4 of Golub and Van Loan [4] has one of the best available discussions of the topic, particularly with regard to roundoff errors.

PROBLEM 1.4

For the two circuits shown:

1. Write Kirchhoff's current law at nodes 1 and 2.
2. Solve for V_2 in terms of V_1 for both KCL equations (at nodes 1 and 2). In the process, the multiplier on V_1 will be the slope in the V_1-V_2 plane.
3. Compare the slopes of the V_2 expressions for both nodes, and do that for both circuits. Which circuit would you expect to find roundoff problems in?
4. Compute the solutions for both networks using three-place arithmetic. Compare your results with the exact solutions.

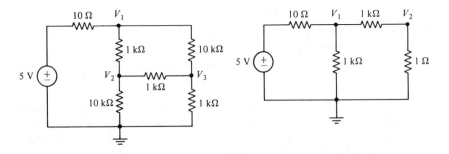

1.7 EXPANDING THE CLASS OF CIRCUITS: CONTROLLED SOURCES

So far we have considered circuits with just resistors and current and voltage sources. We know that there are many other kinds of elements that we have to consider. Energy storage elements (capacitors and inductors) are important components that we will reserve for a later chapter. They introduce dynamic behavior into circuits, and we will look at them in the context of computing transient behavior. However, we should look at the effect of adding controlled sources to the list of permissible elements before we leave this chapter.

Controlled sources are an important class of electrical elements. Controlled sources are found in transistor models, operational-amplifier models, and in any sort of amplifier or active-device model. If a circuit analysis program is going to be able to handle the really "interesting" circuits, it will have to be able to handle controlled sources.

We will begin our consideration of controlled sources with a particular example, the voltage-controlled current source (VCCS). Obviously, later, we will have to worry about all combinations of control (voltage across an element, current through an element) and controlled sources (voltage and current). However, for now we consider just a VCCS embedded in a network as shown in Figure 1.11. We will let

NP = Positive terminal of the controlled source

NM = Negative terminal of the controlled source

V_{NCP} = Voltage of the "noninverting" (positive?) terminal controlling the current source

V_{NCM} = Voltage of the "inverting" (negative?) terminal controlling the current source

G = Transconductance G_m of the element

I = Current in the controlled source = $G(V_{NCP} - V_{NCM})$

Now we can write the node equations at nodes NP and NM.

$$\text{At NP:} \quad \frac{V_{NP} - V_A}{R_A} + G * (V_{NCP} - V_{NCM}) = 0$$

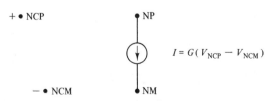

Figure 1.11 A voltage-controlled current source.

$$
\begin{bmatrix} +G_m & -G_m \\ -G_m & +G_m \end{bmatrix}
\begin{array}{l} \text{Row NP} \\[6pt] \text{Row NM} \end{array}
$$

Col NCP   Col NCM

Figure 1.12 Changes in the conductance matrix for a VCCS.

At NM: $\quad \dfrac{V_{\text{NM}} - V_B}{R_B} - G * (V_{\text{NCP}} - V_{\text{NCM}}) = 0$

If we do a linear analysis of this circuit, we need to consider the effects on the conductance matrix G and on the source vector S. In Figure 1.12 we show the entries that have to be made to accommodate the addition of the controlled source.

Each row of the conductance matrix summarizes one node equation, and will usually have two new entries since the controlled current is controlled by the difference in voltage, $V_{\text{NCP}} - V_{\text{NCM}}$. However, if a node turns out to be the ground node (node 0), then that row or column will not exist in the conductance matrix, and fewer new entries will have to be made. When we code this (say, in FORTRAN), we will have to check for those possibilities. Appendix B shows a FORTRAN code that inputs data for a VCCS, and another code that modifies the conductance and source matrix to account for the VCCS.

EXAMPLE 1.4

A linear model of an operational amplifier can consist of just a VCCS with an internal resistance. A 741 operational amplifier connected as an inverter

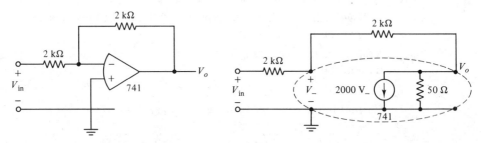

(left) has the linear model shown at the right. Writing KCL at the two nodes indicated, we find

$$
V_- \left[\frac{1}{2000} + \frac{1}{2000} \right] + V_o \left[\frac{-1}{2000} \right] = \frac{V_{\text{in}}}{2000}
$$

$$
V_- \left[\frac{-1}{2000} + 2000 \right] + V_o \left[\frac{1}{2000} + \frac{1}{50} \right] = 0
$$

Thus, the conductance matrix is

$$
G = \begin{bmatrix} 0.0005 + 0.0005 & -0.0005 \\ -0.0005 + 2000 & 0.0005 + 0.02 \end{bmatrix}
$$

The determinant of this system of equations is computed as

$$|G| = 20.5 \times 10^{-6} - [0.25 \times 10^{-6} - 1] \simeq 1.0$$

Interestingly enough, computing the determinant of this system does not involve any sensitive computations even though the circuit has somewhat large ratios of resistance. The large gain of the voltage-controlled current source "swamps" all the other terms in the computation, and the computation is insensitive to any other terms.

PROBLEM 1.5

For the conductance matrix in Example 1.4 determine whether the problem becomes ill conditioned if the gain of the operational amplifier drops to 100 (from 100,000).

PROBLEM 1.6

Determine the conductance matrix for the summing-amplifier operational-amplifier circuit shown and determine whether the problem is ill conditioned.

Once we have gained an understanding of the VCCS, it is easier to see how we can accommodate other types of controlled sources. The next logical step is to consider the current-controlled current source (CCCS).

With a current-controlled current source, we need to know the current flowing through the controlling element before we can calculate the value of the controlled source. However, we will have to calculate that controlling current differently for different elements. If the controlling element is a resistor R, then all we need to know is the voltage difference across the resistor terminals. However, if the controlling element is a voltage source, then the computation of current flowing through the source is a little bit more complicated. If the controlling element is a current source, our task is easy. Those three cases are all that we want to consider for now, and we will take them one by one.

First consider the current-controlled current source, CCCS, with the controlling current flowing through resistor R. Assume that the resistor is connected between nodes NCP (the positive node of the controlling element) and NCM (the negative node). Then we have the situation shown in Figure 1.13.

The expression for the additions to currents into (out of) the nodes NP and NM are essentially the same as for the VCCS, so the additions to the conductance matrix are also very similar, as shown in Figure 1.14.

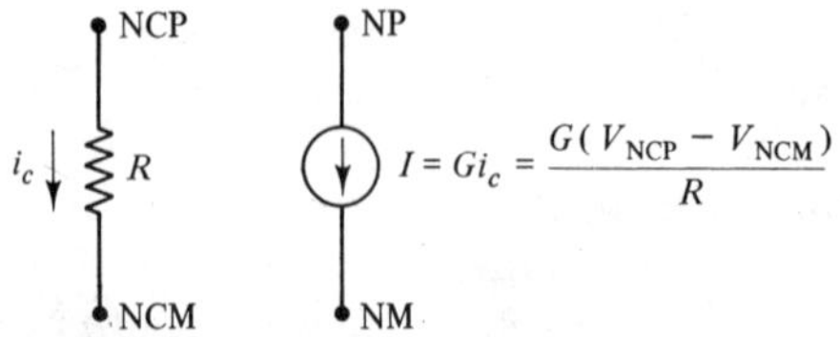

NCP   NP

i_c R $I = Gi_c = \dfrac{G(V_{NCP} - V_{NCM})}{R}$

NCM   NM

Figure 1.13 Current-controlled current source, with controlling current flowing through a resistor.

$$\begin{bmatrix} +G_i/R & -G_i/R \\ -G_i/R & +G_i/R \end{bmatrix} \begin{array}{l} \text{Row NP} \\ \text{Row NM} \end{array}$$

Col NCP   Col NCM

Figure 1.14 Changes in conductance matrix for a CCCS with R control.

Next, if the controlling element is a current source, we will find that we have no effect on the conductance matrix, but that the source vector must be modified. The controlled source is shown in Figure 1.15. The effect on the source vector is as shown in Figure 1.16.

Now, if we have a CCCS controlled by a voltage source (which will include internal resistance in our present formulation) then we have to take the internal resistance of the source into account. This situation can also be diagrammed, as in Figure 1.17. A current source controlled by a voltage source with internal resistance will affect both the conductance matrix and source vector. The situation is shown in Figure 1.18.

This accounts for all controlled current sources. We still need to be able to handle controlled voltage sources. However, all of our voltage sources in this formulation have nonzero internal resistance, so we can do a voltage source to current-source transformation. That will yield almost the same situation as we faced with the controlled current sources, except that each controlled source will have an internal conductance that has to be accounted for. Taking care of that internal conductance is only a minor problem, since it is handled like any other conductance.

With just these three controlling elements (R, E, and I) we can cover a multitude of situations. Whenever we consider circuits with inductors, capacitors, and diodes, we will use methods of analysis that model those elements with R,

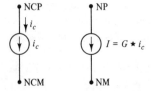

NCP   NP

i_c

i_c $I = G \star i_c$

NCM   NM

Figure 1.15 Current-controlled current source with current-source control.

$$\begin{bmatrix} G \times J \\ -G \times J \end{bmatrix} \begin{array}{l} \text{Row NP} \\ \text{Row NM} \end{array}$$

Figure 1.16 Changes in source vector for a CCCS with current control.

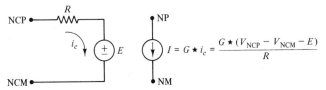

Figure 1.17 Current-controlled current source with voltage-source control.

$$
\begin{bmatrix} +G/R_{int} & -G/R_{int} \\ -G/R_{int} & +G/R_{int} \end{bmatrix} \quad \begin{matrix} \text{Row NP} \\ \text{Row NM} \end{matrix} \quad \begin{bmatrix} V_s\,G/R_{int} \\ -V_s\,G/R_{int} \end{bmatrix}
$$

Col NP   Col NC

Conductance Matrix   Source Vector

Figure 1.18 Changes for CCCS with voltage control.

E, or I models (for transient analysis and for solution of nonlinear circuit problems).

PROBLEM 1.7

Example 1.4 discussed an operational amplifier circuit using a voltage-controlled current source. Redo that analysis using a voltage-controlled voltage source as pictured. Are conclusions about sensitivity and ill conditioning changed as a result of your analysis?

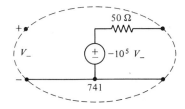

PROBLEM 1.8

Using a voltage-controlled voltage source model construct the conductance matrix for the circuit shown and determine whether the solution is sensitive to circuit parameters near the parameters given.

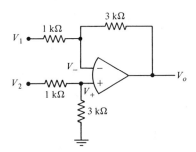

PROBLEM 1.9

In the circuit shown, determine the value of gain in the voltage-controlled voltage source for which the solution becomes indeterminate.

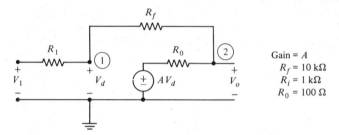

Gain = A
R_f = 10 kΩ
R_i = 1 kΩ
R_0 = 100 Ω

1.8 A SUMMARY AND A PREVIEW

At this point we have had an introduction to one way of doing circuit analysis numerically, and we have started to look at some of the problems that can arise in this type of analysis. The DC circuit analysis programs that we have considered can form a base from which we can explore other topics.

In succeeding chapters we will expand on this introductory material. First, we will look at the problem of how to incorporate some nonlinear circuit elements. Then, later, we will add energy storage elements. In both cases we will try to do things in a way that will eventually produce a working circuit analysis program. Our efforts to this point are put together in Appendix A, where we will find a simple DC circuit analysis program using the programs and fragments developed in this chapter. In the next chapter we will modify our programs to include diodes and other nonlinear elements. Later, we will go on to include inductors and capacitors. Our work there will prompt examination of ways to do numerical integration of the state equations describing the dynamics of these networks. We will also look at numerical formulation of state equations and optimization of network functions.

REFERENCES

1. R. W. Hamming, *Numerical Methods for Scientists and Engineers,* McGraw-Hill, New York, 1973.
2. D. F. Tuttle, *Circuits,* McGraw-Hill, New York, 1977.
3. A. Ralston and P. Rabinowitz, *A First Course in Numerical Analysis,* 2nd ed., McGraw-Hill, New York, 1978. (The quote on complex matrices is found on page 465, in a short discourse on "Miscellaneous Topics." The entire book is worth reading and is invaluable to anyone interested in numerical analysis.)
4. G. H. Golub and C. F. Van Loan, *Matrix Computations,* Johns Hopkins University Press, Baltimore, 1983.
5. A. S. Deif, *Advanced Matrix Theory for Scientists and Engineers,* Halsted Press/ Wiley, New York, 1982.
6. S. W. Director and R. A. Rohrer, *Introduction to System Theory,* McGraw-Hill, New York, 1972.

MINIPROJECTS

1. Section 1.6 presented a particular circuit that could present difficulties in the matrix inversion solution. On your computer, determine the lower limit of resistance for the present 1-Ω resistor for which good numerical solutions can be obtained. If you have a circuit analysis package available (for example, SPICE) do the same for the circuit package.

2. We mentioned early in this chapter that there were usually many different algorithms that could be used to accomplish any numerical task. Matrix inversion falls in that class.

   An alternative method for calculating the inverse of a matrix relies on the Cayley-Hamilton theorem. This theorem is usually paraphrased (much to the general disgust of true mathematicians everywhere) by saying that a matrix satisfies its own characteristic equation. (See Deif [5] for a statement of this theorem, or Director and Roherer [6] for a readable proof.) The theorem says

$$p_0 I + p_1 A + p_2 A^2 + \cdots + p_n A^n = 0 \qquad A \text{ is } n \times n$$

and

$$|\lambda I - A| = p_0 + p_1 \lambda + p_2 \lambda^2 + \cdots p_n \lambda^n$$

If we premultiply by the inverse of A, we find that we can solve for the inverse of A:

$$A^{-1} = \frac{-1}{p_0} [p_1 I + p_2 A + \cdots + p_n A^{n-1}]$$

The question-raised here is whether or not this is a reasonable way to compute an inverse. *Assume* that you have the polynomial coefficients, and from that point on determine how many operations are needed to compute the inverse. If A is $n \times n$, express your results as a function of n.

Chapter 2

Numerical Solutions for Nonlinear Networks

2.1 AN INTRODUCTION TO NONLINEAR ANALYSIS

Nonlinear circuits are the most interesting and most generally useful circuits. Linear circuits are often used to filter out noise or signals within a certain frequency range. However, only nonlinear circuits can introduce new frequencies into a signal (whether wanted or not). Only nonlinear circuits exhibit the peculiar behavior we take advantage of when we modulate a signal or detect a modulated signal.

The analysis of nonlinear networks produces special problems. Our favorite signals, such as damped sinusoids, exponentials, and the like are the kinds of natural responses found in linear circuits. Nonlinear circuits exhibit such a wide range of peculiar behaviors that it is hard to catalog them [1–3]. Worse yet, there are no general solution methods that work for a wide range of nonlinear problems; so when we try to analyze these networks numerically, we enter an area populated more by ad hoc ideas, often lacking firm theoretical underpinnings of the sort we have for linear circuits and systems. We will find that it is easy to pose problems with miserable numerical properties—or even impossible of solution. In this chapter we will move through a sequence of problems, going from simple to complex. First, however, we examine a simple, but miserable problem—one that illustrates the perils that await us in this area.

EXAMPLE 2.1

In the circuit shown, the nonlinear device has a characteristic that is described by the following quadratic polynomial:

$$v = 2i(1 - i) = 2i - 2i^2$$
$$v_s = 6 \text{ V} \qquad R = 3 \,\Omega$$

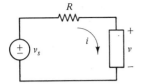

The voltage at the terminal of the source is $(v_s - iR)$ and is equated to the voltage across the nonlinear element, solving eventually for the current through the element.

$$6 - 3i = 2i - 2i^2$$
$$2i^2 - 5i + 6 = 0$$
$$i = \frac{5 \pm \sqrt{25 - 4 \times 12}}{4} = 1.25 \pm \frac{\sqrt{-23}}{2}$$

This current is complex. There can be no such thing as a complex current, and hence there is no solution that exists for the network we have described.

This example illustrates that it is possible to describe a network mathematically in a way that produces a situation with no solution, even when we may be convinced that the description is valid. (However, it must be that this model overlooks something in the physical situation that actually determines the solution.)

The question of existence of solutions has been a subject of interest over the years. References [2]–[4] are reasonable starting points for reading in this area. In the remainder of this chapter we will be assuming that our mathematical description of a network has been produced with an eye to this problem. Nevertheless, any network analysis program will have to be cognizant of the problem.

PROBLEM 2.1
The popular circuit analysis program SPICE is described in Appendix D. Write and run a SPICE program to evaluate the circuit in Example 2.1. Comment on the results.

2.2 AN EXAMPLE NETWORK

We will start by examining the example network shown in Figure 2.1. The nonlinear element is described mathematically by

$$i = 0.1v^{0.6} \qquad\qquad (2.2.1)$$

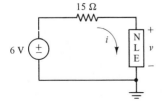

Figure 2.1 A simple circuit containing a nonlinear element.

This nonlinear element is a representative of the general class we want to consider. It is a zero memory nonlinear element; that is, the current i depends upon the value of the voltage v at the present instant. Past values of voltage have no effect upon the present current. Since the element doesn't "remember" past voltages, it is referred to as a zero memory element.

We will examine the problem of obtaining an analytic solution for the voltage and current in this network. If we write Kirchhoff's current law at the output node for this circuit, we have

$$\frac{v-6}{15} + 0.1v^{0.6} = 0 \qquad (2.2.2)$$

Now we could attempt to get an analytic solution to this equation. Try it. You won't have much luck solving this equation, even though you might recast it in ways that make you think you are making progress. However, no matter what you do, you will not be able to generate an analytic solution to this equation, and it is impossible to get one. (You might eventually get the equation reformulated as a polynomial with integral exponents, but it will be of a degree higher than it is theoretically possible to solve.)

In this situation, the only way a solution can be obtained is to generate a numerical solution, using some sort of algorithm that can be programmed. Almost all numerical algorithms start with some initial estimate of the solution, and iterate somehow in an effort to converge to the correct solution. The simplest kind is a search algorithm, which we will consider first.

2.3 A SIMPLE SEARCH ALGORITHM

If we pick a voltage value at random, and evaluate the left-hand side of the Kirchhoff current law expression, that expression will be either positive, negative, or zero. If it is zero, then we have gotten very lucky and guessed the value of voltage in the network, and we can go on to compute the current using Eq. (2.2.1). However, chances are that we won't be that lucky. On the other hand, if we can guess a few more values and find two values of voltage, one of which produces a positive value in Eq. (2.2.2) and the other gives a negative value, then we know that the solution to Eq. (2.2.2) lies between these two values. Once we have the solution bracketed, we can evaluate Eq. (2.2.2) in the middle of the bracketed interval, and, depending on whether the value of the result is positive or negative, we can eliminate one or the other half of the original interval.

Figure 2.2 shows how such a search would proceed. In this particular in-

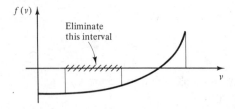

Figure 2.2 A binary search.

stance, we have some information that we can use, since we know that the voltage has to lie somewhere between zero volts (ground) and the supply voltage, +6 volts. If we define the function, $f(v)$:

$$f(v) = 0.1v^{0.6} + \frac{v-6}{15} \tag{2.3.1}$$

Evaluating $f(v)$ at the two points, $v = 0$ and $v = 6$, we find

$$f(0) = -\frac{6}{15} < 0 \tag{2.3.2}$$

and

$$f(6) = (0.1)^{0.6} > 0 \tag{2.3.3}$$

So we can conclude that the solution lies somewhere between these two extremes. If we evaluate the function at $v - 3$, we find

$$f(3) = \frac{3-6}{15} + 0.1(3^{0.6})$$
$$= -0.2 + 0.1933182 < 0 \tag{2.3.4}$$

From this, we have to conclude that the solution does not lie in the interval from 0 to 3 V, but that it does lie in the interval from 3 to 6 V. We can continue in this manner, halving the interval as we go along, homing in on the solution. We may never get the precise solution, but we can get as close to it as we want.

ALGORITHM

BINARY SEARCH

The binary search algorithm can be used to determine the location of a zero of a function of a single variable. In the binary search algorithm the function is evaluated at the endpoints and the midpoint of an interval, and one half of the interval is discarded if the function is of the same sign at both ends of the half-interval.

1. Evaluate $F_{left} = F(X_{left})$.
2. Evaluate $F_{right} = F(X_{right})$, where X_{right} = rightmost point.
3. Evaluate $F_{mid} = F(X_{mid})$, where X_{mid} = middle point.
4. If F_{right} and F_{mid} have the same sign, discard the right half of the interval from X_{mid} to X_{right}, and continue the search from X_{left} to X_{mid}, unless sufficient accuracy has been achieved.
5. If F_{left} and F_{mid} have the same sign, discard the left half of the interval from X_{left} to X_{mid}, and continue the search from X_{mid} to X_{right}, unless sufficient accuracy has been achieved.

PROBLEM 2.2

Manually carry the iterative search along until the solution is within 0.05 V of the true solution.

PROBLEM 2.3

Write a program to do the iterative search, and calculate the voltage to within 0.1 V.

PROBLEM 2.4

In this diode circuit we can write an expression for the KCL "mismatch" as shown. The output voltage should be somewhere between 5 and 6 V.

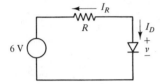

Search the range between 5 and 6 V using intervals of 0.1 V, and comment on the (un)symmetrical nature of the mismatch on either side of the solution.

$$f(v) = I_0(e^{qv/kT} - 1) + \frac{v-6}{R}$$
$$I_0 = 10^{-9}\,a \qquad -q/kT = 40 \qquad R = 15\ \Omega$$

2.4 THE FUNCTIONAL ITERATION ALGORITHM

Search methods of the sort presented in Section 2.3 are sometimes good. However, they do have some drawbacks. The most serious drawback is that the rate of convergence is not the best that can be obtained. There is also going to be a problem trying to extend this method to more general situations in which we have to search many different node voltages. (If instead of finding differences in sign at ends of an interval, we found differences in sign at the corners of a square in a v_1-v_2 plane, what would that mean?) So, even though search methods look like they will always converge, it will be worthwhile looking at several other algorithms. In the next section we will look at the popular Newton-Raphson algorithm. Before that, however, we will look at a very simple algorithm that can sometimes do well. It makes a good starting point for our discussion of methods for solving nonlinear problems.

The first algorithm we will look at is so simple that it is deceiving. We start by rewriting the equation we are going to solve. Earlier, we had

$$\frac{v-6}{15} + 0.1v^{0.6} = 0 \tag{2.4.1}$$

We can rewrite this equation to "solve" for v, as follows:

$$v = (-15)(0.1v^{0.6}) + 6 = g(v) \tag{2.4.2}$$

Of course, v still appears on both sides of this equation, and we haven't really solved for anything. Nevertheless, if we have a value for v, we could insert our value into the right-hand side of this expression, and get a new value of v. It is anything but obvious that this will be worth doing, but let us try anyway. In Appendix C we have a program that does just that (for general source voltages and resistance). The results are given in Table 2.1.

These results look very interesting. It seems clear that the computed values are converging to a value near 3.06+. This algorithm can be generalized easily, and it is used on occasion. It gives us a benchmark that we can use when we look at other algorithms applied to this same circuit.

This algorithm can take us into many interesting side issues. First of all, the reason the algorithm works is that the function we defined, $g(v)$, is a "contraction." A contraction always lessens the distance when two values of v are mapped. What we mean is:

$$|g(v_a) - g(v_b)| \leq B|v_a - v_b| \quad \text{where} \quad B < 1 \quad (2.4.3)$$

TABLE 2.1

```
TYPE SOURCE VOLTAGE AND INITIAL GUESS FOR V.
?6.,6.,

TYPE THE NUMBER OF ITERATIONS.
?20,

TYPE THE VALUE OF THE RESISTOR (OHMS).
?15.,
```

ITERATION	VOLTAGE
1	1.604766
2	4.007776
3	2.549887
4	3.369709
5	2.890814
6	3.164019
7	3.006118
8	3.096680
9	3.044513
10	3.074488
11	3.057240
12	3.067156
13	3.061452
14	3.064732
15	3.062845
16	3.063931
17	3.063306
18	3.063665
19	3.063459
20	3.063578

for v_a and v_b within some predetermined region. If a function is a contraction, then it is known that:

1. There exists a v' such that

$$v' = g(v') \tag{2.4.4}$$

   (That is, a solution exists.)
2. A sequence of iterates, $v(k)$, will converge to the solution v' when:

$$v(k+1) = g(v(k)) \tag{2.4.5}$$

3. For some $N \geq |g(v_0) - v_0|$ $\qquad\qquad\qquad\qquad\qquad\qquad$ (2.4.6)

$$|v(k) - v_0| \leq \frac{NB^k}{1-B} \tag{2.4.7}$$

Thus, if a function is a contraction this process is guaranteed to converge to the unique solution. Moreover, we know something about the rate of convergence. Finally, it has to be pointed out that a function is a contraction if it has a derivative less than one!

Our function, $g(v)$, doesn't have a derivative less than one everywhere, but for some region near the solution, it does satisfy that condition.

This method seems novel and cute, but it is not a method that has any obvious a priori justification. Still, there are many applications where it has been used. Chua and Lin [5] present the functional iteration method as a viable circuit calculation technique, so it does seem to get used. However, Hofstadter [6] shows how even very simple functions can produce literal chaos when iterated in this manner.

ALGORITHM

CONTRACTION MAPPING

The contraction mapping algorithm can sometimes be used to find the fixed point of a function, that is where $f(x) = x$. Many zero-finding problems can be recast in such a form. The algorithms steps are as follows:

1. Determine a starting value, x_0.
2. Evaluate $f(x_0)$.
3. Let $x_1 = f(x_0)$ and then evaluate $f(x_1)$.
4. Continue iterating, using the last found value of $f(x)$ as the value of x in the function.
5. Iterate until either sufficient accuracy is achieved or until some predetermined number of iterations has been done.

2.5 THE NEWTON-RAPHSON ALGORITHM

The search algorithm and the iterative function computation look as though both would be workable methods for many problems. However, they may con-

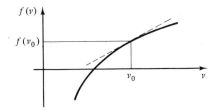

Figure 2.3 A function and a linear extrapolation around v_0.

verge too slowly for many applications. In this section we will examine the popular Newton-Raphson algorithm, which exhibits good convergence—one we can compare to the methods developed in the preceding two sections. It is easily generalized to cases in which less information is available, and it has a "neat" interpretation in terms of circuit variables.

To see what this algorithm involves, let us examine a case where we are trying to compute the root of a function. Say we have a graph of the function. We do that just to visualize what goes on in the algorithm, as in Figure 2.3.

We will assume that we have some initial guess at the solution. Call that guess v_0. To compute the next, improved (we hope) estimate of the root, we do a linear extrapolation around our initial guess, and solve for the point at which the linear extrapolation becomes zero. We then take that as our next estimate for the location of the root. This is shown graphically in Figure 2.4.

Analytically, we accomplish what is illustrated in the graph by developing a linear approximation, $f_{lin}(v)$, around the most recent guess at a solution; that is,

$$f_{lin}(v_1) = f(v_0) + f'(v_0)(v_1 - v_0) \tag{2.5.1}$$

Setting $f_{lin}(v_1)$ equal to zero and solving, we obtain

$$v_1 = v_0 - \frac{f(v_0)}{f'(v_0)} \tag{2.5.2}$$

Continuing in this way, we can generate better and better estimates of the root of the equation, by taking v_1 as the starting point for our next iteration. This process is not entirely foolproof, but chances of convergence are good. There does exist, however, the possibility of a situation of the sort shown in Figure 2.5.

Now, examine how to apply this algorithm to our example network. From Kirchhoff's current law, we had

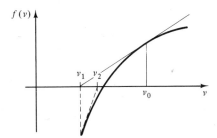

Figure 2.4 The first step using a linear extrapolation.

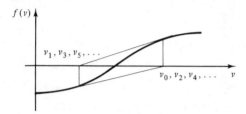

Figure 2.5 The algorithm caught in a never-ending cycle.

$$f(v) = \frac{v-6}{15} + 0.1v^{0.6} = 0 \tag{2.5.3}$$

Applying Newton's algorithm, we would have

$$v_1 = v_0 - \frac{f(v_0)}{f'(v_0)} \tag{2.5.4}$$

This algorithm is easily programmed, and a simple program is given in Appendix C. Running that program gives the results in Table 2.2.

Note the rapid convergence with this algorithm. Compare how rapidly this method converges with the rate of convergence of the function iteration (data in Table 2.1).

PROBLEM 2.5

Run the program with a starting value of 1000 V. Comment on the rapidity of the convergence. Is it rapid? Does there seem to be any pattern to the distance from the final convergence point?

TABLE 2.2

```
 TYPE SOURCE VOLTAGE AND INITIAL GUESS FOR V.
?6.,6.,

 TYPE THE NUMBER OF ITERATIONS.
?5,

 TYPE THE VALUE OF THE RESISTOR (OHMS).
?15.,

 ITERATION      VOLTAGE

 1              2.946744
 2              3.063199
 3              3.063534
 4              3.063534
 5              3.063534
 *EXIT*
```

ALGORITHM

NEWTON-RAPHSON

The Newton-Raphson algorithm can be used to find the root of a function $f(x)$. The algorithm has but a single computational step:

Iterate until either a specified accuracy or number of iterations has been performed with the iterative formula:

$$x_{k+1} = x_k - \frac{f(x_k)}{f'(x_k)} \qquad k = 0, 1, \ldots ?$$

ALGORITHM

DERIVATIVE COMPUTATION

There will be instances in which a function form is available, but no analytic form for the derivative is available. (In the example data presented in Table 2.2 the derivative functional form was used directly.) When the derivative is not given explicitly, it will be necessary to compute the derivative numerically. When that happens, some attention should be paid to the accuracy of the computation.

The most obvious way to compute a derivative is simply to do a finite-difference approximation. Let us examine that process representing the function $f(x)$ with several terms in a Taylor series.

$$f(x + \Delta x) = f(x) + \frac{df}{dx}\bigg|_x \Delta x + \frac{d^2f}{dx^2}\bigg|_x \frac{(\Delta x)^2}{2} + \frac{d^3f}{dx^3}\bigg|_x \frac{(\Delta x)^3}{6} + \cdots$$

$$D_+[f(x)] = \frac{f(x + \Delta x) - f(x)}{\Delta x}$$

$$= \frac{df}{dx}\bigg|_x + \frac{d^2f}{dx^2}\bigg|_x \Delta x + \frac{d^3f}{dx^3}\bigg|_x (\Delta x)^2 + \cdots$$

$$\text{Error} \simeq \frac{d^2f}{dx^2}\bigg|_x \frac{\Delta x}{2} + \cdots$$

Obviously, the error is proportional to the size of the increment that is used.

Now, consider a slightly different finite-difference method, and once again compute the error.

$$D_s[f(x)] = \frac{f(x + \Delta x) - f(x - \Delta x)}{2 \Delta x}$$

$$= \frac{df}{dx}\bigg|_x + \frac{d^3f}{dx^3}\bigg|_x (\Delta x)^2 + \cdots$$

$$\text{Error} \simeq \frac{d^3f}{dx^3}\bigg|_x (\Delta x)^2$$

Here the error is proportional to the square of the increment. If the increment is small, then this method can be much more accurate than the forward increment method.

PROBLEM 2.6

Numerically evaluate the derivative of the cubic function around the points 1, 2, and 5. Use increments of 0.01 and 0.001. Compare your results with the claims above.

$$f(x) = x^3$$

PROBLEM 2.7

Rewrite the program in Appendix C so that it uses a numerical computation of the derivative in the Newton algorithm. Compare the effects of using the forward-difference method, and the symmetric difference method for computing the derivative.

A finite-difference approximation to a derivative has another problem that can arise. In taking finite differences, it is possible that roundoff will affect the computed difference. We might be tempted to think that taking the increment between function evaluations smaller and smaller will always increase accuracy. That is not so. Sooner or later, the finite wordlength of the computer will cause both function values to be equal. The difference between them is smaller than can be represented in the machine. When that happens, the computed difference is zero, and the computed derivative is also zero.

PROBLEM 2.8

For the cube function, determine the smallest increment that produces a zero computed derivative for values of 1, 2, and 5 using:

$$\text{(a)} \quad D_+[f(x)] = \frac{f(x + \Delta x) - f(x)}{\Delta x}$$

$$\text{(b)} \quad D_s[f(x)] = \frac{f(x + \Delta x) - f(x - \Delta x)}{2\Delta x}$$

Before going further, we should note some potential problems with the programming approach we used here. We did not have any rule for stopping the algorithm. We did a predetermined number of iterations, and accepted the result that was computed in that predetermined number of iterations. A better approach would be to stop whenever we achieved acceptable accuracy. Since we wouldn't necessarily want the user to make that decision every time, a stopping rule should be built into our program. However, we also need to do that knowledgeably; that is, we should know something about how rapidly we should expect the method to converge. We will start to explore that topic in the next section.

POINTS TO PONDER: In the Newton-Raphson algorithm we construct a linear model that is valid only locally around a point. Then we find a solution using the linear model in place of the actual nonlinearity. There is no conceptual problem in generating a quadratic model and using it much the way we use the linear model. What is involved if we use a quadratic model? Outline an algorithm in detail using a quadratic model.

2.6 CONVERGENCE PROPERTIES OF THE NEWTON-RAPHSON ALGORITHM

One important reason for considering different algorithms for solution of non-linear equations is that different algorithms have differing convergence rates. A straightforward search algorithm will examine points throughout the search region, and it may spend time evaluating points in regions where there is little possibility of finding the solution. It would make sense to try to get a "smarter" algorithm, and the Newton-Raphson algorithm may be smarter in that sense since it tries to home in on a solution and refine the accuracy of the computed solution. It is worthwhile looking at the convergence properties of an algorithm and in this section we will look at the Newton algorithm in some detail. The development in this section will parallel that in Daniels [7]. We will get an approximate expression for the error at each stage of the computation, and determine how that error varies as the computation proceeds. A more mathematically precise and rigorous approach can be found in Chapter 2 of Dennis and Schnabel [8]. However, let's get on with our development.

Let us assume that we are trying to find the root of a function, $f(x)$. Expand the function $f(x)$ around the root we are trying to find. (We can do that in principle, even though we do not know where that root might be.) Imagine that the root is located at $x = r$. Take the expansion out as far as the quadratic term.

$$f(x) = f(r) + (x - r)f'(r) + \frac{1}{2}(x - r)^2 f''(r) \qquad (2.6.1)$$

where $f(r) = 0$, since r is a root of $f(x)$.

We can get an analytic expression for $f'(x)$ and plug that expression into the iteration formula for the Newton algorithm.

$$f'(x) = f'(r) + (x - r)f''(r) \qquad (2.6.2)$$

So

$$x_{k+1} = x_k - \frac{f(x_k)}{f'(x_k)} \qquad (2.6.3)$$

Next, we examine the error made at step $(k + 1)$.

$$E_{k+1} = x_{k+1} - r \qquad (2.6.4)$$

$$= x_k - r - \frac{f(x_k)}{f'(x_k)} \tag{2.6.5}$$

$$= E_k - \frac{f(r) + E_k f'(r) + E_k^2 f''(r)/2}{f'(r) + E_k f''(r)} \tag{2.6.6}$$

$$= E_k \left[1 - \frac{(1 + (E_k/2)(f''(r)/f'(r)))}{(1 + E_k(f''(r)/f'(r)))} \right] \tag{2.6.7}$$

This is approximately

$$E_{k+1} \simeq E_k \frac{(E_k/2)f''(r)}{f'(r)} \tag{2.6.8}$$

Thus, the error at step $(k + 1)$ is approximately proportional to the square of the error at step k, and this process is said to be quadratically convergent.

PROBLEM 2.9

Check the development just presented against the data in Table 2.1. Does the algorithm converge quadratically as claimed above? If not, does it converge faster or slower than the quadratic rate discussed above?

Convergent Series

Frequently we do computations that involve a series of terms, and we are interested in the way those kinds of computations converge. For example, we are all familiar with power series expansions for things like exponential functions and trigonometric functions. In other cases, a computation might be done iteratively in a way that causes the computational results to converge to an answer (and that computation might not necessarily involve summing a series).

Whenever we deal with a sequence of computational results, the most important item of interest is whether or not the sequence converges. However, the rate at which the sequence converges might be just as important. If a computation converges so slowly that it takes excessive CPU time (more than our budget, for example) then we would not be interested in that method of computation. In almost all cases we can think of, we would also be interested in speeding up a computation as long as we do not suffer a penalty in accuracy.

It is probably best if we consider a few examples of the kind of process we are talking about. The first example, implemented in FORTRAN, is a functional iteration:

```
C
C PROGRAM TO ILLUSTRATE FUNCTION ITERATION CONVERGENCE
C
      X = .1
      ALPHA = 1/.51
      WRITE (6,*)' Iteration #, X, (X-.49)'
      I = 0
      WRITE (6,*) I,X, (X-.49)
```

```
    DO 100 I = 1,10
       X = ALPHA*X*(1.-X)
       WRITE (6,*) I,X, (X-.49)
100 CONTINUE
    CALL EXIT
    END
```

This program produces the results:

```
Iteration #, X, (X-.49)
0   .1000000  -.3900000
1   .1764706  -.3135294
2   .2849583  -.2050417
3   .3995236  -.9047637E-01
4   .4704010  -.1959903E-01
5   .4884782  -.1521774E-02
6   .4899358  -.6422400E-04
7   .4899975  -.2533197E-05
8   .4899999  -.1043081E-06
9   .4900000  -.7450581E-08
10  .4900000  -.7450581E-08
*EXIT*
```

This seems to be a fairly rapidly converging series. Now consider another program.

```
C
C PROGRAM TO ILLUSTRATE CONVERGENCE OF HIGHER ORDER
C
    X = .1
    WRITE (6,*)' Iteration #, X, (X-.49)'
    I = 0
    WRITE (6,*) I,X, (X-.49)
    DO 100 I = 1,10
       X = .5*(X + (.2401/X))
       WRITE (6,*) I,X, (X-.49)
100 CONTINUE
    CALL EXIT
    END
```

Again, we can look at the results of running the program. In this program, the computed results converge to the same point, 0.49, starting in both cases from 0.1. However, the second process seems much more rapidly convergent.

```
Iteration #, X, (X-.49)
0   .1000000  -.3900000
1   1.250500   .7605000
2   .7212516   .2312516
3   .5270726   .3707255E-01
4   .4913038   .1303770E-02
5   .4900017   .1721084E-05
```

```
 6  .4900000   -.7450581E-08
 7  .4900000   -.7450581E-08
 8  .4900000   -.7450581E-08
 9  .4900000   -.7450581E-08
10   .4900000   -.7450581E-08
*EXIT*
```

This second process seems to have the property that it converges very rapidly when it gets close to the ultimate value.

The important concept here is that there is a difference in the rate of convergence of these two processes. Even though both processes start from 0.1 and end up at 0.49, the path taken to get to the limit is not the same. Clearly we should take the fastest path, and we need to know something about what determines how fast the convergence is.

For the first program (functional iteration) the ratio of successive errors (the term $x - 0.49$) approaches something like 0.04 (actually 0.0392156 if the process is carried out with infinite precision and many times). In that kind of situation, the error at any iteration is proportional to the error found in the previous iteration.

For the second program, the error in any iteration is proportional to the square of the error in the previous step. In this case, when we get to an iteration with a small error, squaring the error dramatically reduces the error in the next step.

Let E_k be the error at the kth step. Then the first process is said to converge "linearly" since E_{k+1} is linearly proportional to the previous error.

$$E_{k+1} = \text{Constant} \times E_k$$

The second program is said to converge quadratically since E_{k+1} is proportional to the square of the previous error.

$$E_{k+1} = \text{Constant} \times E_k \times E_k$$

Clearly, quadratic convergence is preferable to linear convergence, even in this case where the linear constant has a value of 0.04. (However, what if the constant were 0.98?)

This note on convergence rate just touches the surface. For a good introductory piece on convergence see pages 56–58 of Gill, Murray and Wright [9].

PROBLEM 2.10

Calculating the square root of a number is a problem that can be solved using the Newton-Raphson algorithm. Assume that we want to take the square root of a number, a. Then form the function

$$f(x) = x^2 - a$$

To find the zero of this function (the square root of "a") we use the Newton-Raphson algorithm:

$$x_{k+1} = x_k - \frac{f(x_k)}{f'(x_k)} = x_k - \frac{x_k^2 - a}{2x_k}$$

Using the Newton-Raphson algorithm:

1. Calculate the square root of 10, using 3 as a starting value since it is the square root of 9 (close to 10).
2. Determine whether the convergence is quadratic, as claimed earlier.
3. Determine whether the convergence rate constant is correctly given by $f''(r)/2f'(r)$.

2.7 CONVERGENCE RATE IN A SIMPLE DIODE CIRCUIT

In Section 2.6 we indicated that the Newton algorithm should converge quadratically. In many instances that is true. However, the derivation given assumed that we were close to the ultimate solution. The approximations made may not work well if we are far away from the solution. In this section we will examine this problem by working through the solution of a diode-resistor circuit with Newton's algorithm.

Earlier in this chapter we considered a simple circuit with just one nonlinear device. Now we will replace that nonlinear device with a semiconductor diode. Figure 2.6 shows the circuit we will use.

We are going to assume a particular mathematical model for the diode. The simplest model for a diode is [9]

$$i = I_0(e^{qv/kT} - 1) \tag{2.7.1}$$

where

$$q = \text{Electronic charge}$$
$$k = \text{Boltzmann's constant}$$
$$T = \text{Absolute temperature (kelvins)}$$
$$I_0 = \text{Reverse saturation current}$$

It isn't difficult to change our program to account for this different nonlinearity, and the changed program is listed in Appendix C. A run using that program is given in Table 2.3.

There seems to be a problem here. This process is converging very slowly. The change in the computed voltage is the same each time (−0.025 V). Something peculiar seems to be going on.

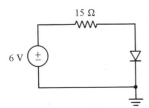

Figure 2.6 A simple diode circuit.

TABLE 2.3

```
TYPE SOURCE VOLTAGE AND INITIAL GUESS FOR V.
?6., 3.,

TYPE THE NUMBER OF ITERATIONS.
?20,

TYPE THE VALUE OF THE RESISTOR (OHMS).
?15.,

ITERATION       VOLTAGE

1               2.975000
2               2.950000
3               2.925000
4               2.900000
5               2.875000
6               2.849999
7               2.824999
8               2.799999
9               2.774999
10              2.749999
11              2.724999
12              2.699999
13              2.674999
14              2.649999
15              2.624999
16              2.599998
17              2.574998
18              2.549998
19              2.524998
20              2.499998
```

Let's look at the details of what is happening here. At each interaction we have

$$v_{n+1} = v_n - \frac{[(v_n - v_s)/R + I_0(e^{qv_n/kT} - 1)]}{[(qI_0/kT)e^{qv_n/kT} - 1/R]} \qquad (2.7.2)$$

If the voltage is large enough, the exponential will dominate in both the numerator and denominator of the increment. Then, the iteration will approximate

$$v_{n+1} = v_n - \frac{kT}{q} \qquad (2.7.3)$$

In our program we used a value of 40 for q/kT, so an increment of -0.025 is just what we should have expected. The algorithm will continue like this until the exponential ceases to dominate. Then we might get quicker convergence as the process gets close to the solution.

PROBLEM 2.11

Run the diode program and check whether the convergence becomes quadratic or faster in any way as the program gets close to the solution. Find the solution.

Obviously we have some problems here. If we use the Newton algorithm in diode circuits (and transistor circuits as well), then we have reason to be worried about the rate of convergence. Either we need to change something to improve convergence rate, or we need to be sure that we start close to the solution every time. Otherwise, we could expend a lot of CPU time.

In a practical circuit analysis program, conservation of CPU time is of some concern, and we cannot avoid analyzing circuits with diodes and transistors (both with exponential model forms). Consequently, in any practical program we will probably have to take pains to ensure that we start close enough to a solution that it converges rapidly, and we would want to avoid situations like the example where we get constant, small increments in our computed solution.

There is a danger here, because behavior of the algorithm is not only poor far from the solution, but the behavior is not exactly symmetric. If we start at a voltage value below the solution, then there can be peculiar behavior. Table 2.4 shows what happens when we take 0 V as our starting point. At zero voltage the diode looks much like an open circuit, so in the first iteration, we calculate almost open-circuit voltage across the diode, then begin the slow descent to the solution from the open-circuit voltage. Clearly, this is also an undesirable sequence of events, and we will need to avoid this kind of behavior in any realistic program. One way to overcome the difficulty is to ensure that the starting "guesses" for diode voltages are always close to a solution. Doing that can produce the quadratic convergence we hope to have in the Newton method.

2.8. INTERPRETING THE NEWTON-RAPHSON ALGORITHM IN CIRCUIT TERMS

In the preceding section we looked at the Newton-Raphson algorithm and discussed a few of the problems associated with it. It will help us somewhat if we can reexamine that algorithm in a different light. Doing that might give some insight into how to change things to get different convergence properties. In this section we will interpret the algorithm in circuit terms, hoping to gain more insight.

We start with our original expression for our original circuit from Section 2.5. The update algorithm at the first step is

$$v_1 = v_0 - \frac{f(v_0)}{f'(v_0)} \tag{2.8.1}$$

Working from this expression, we can devise a circuit interpretation of the algorithm.

TABLE 2.4

```
TYPE SOURCE VOLTAGE AND INITIAL GUESS FOR V.
?6., 0.,

TYPE THE NUMBER OF ITERATIONS.
?20,

TYPE THE VALUE OF THE RESISTOR (OHMS).
?15.,

ITERATION      VOLTAGE

1              5.999997
2              5.974997
3              5.949997
4              5.924997
5              5.899997
6              5.874997
7              5.849996
8              5.824996
9              5.799996
10             5.774996
11             5.749996
12             5.724996
13             5.699996
14             5.674996
15             5.649996
16             5.624996
17             5.599995
18             5.574995
19             5.549995
20             5.524995
*EXIT*
```

First we rearrange Eq. (2.8.1) to obtain

$$f'(v_0) \times (v_1 - v_0) = -f(v_0) \tag{2.8.2}$$

Expand the function and its derivative to get

$$\left(\frac{1}{15} + 0.06v_0^{-0.4}\right)(v_1 - v_0) = \frac{-(v_0 - 6)}{15} - 0.1v_0^{0.6} \tag{2.8.3}$$

In this equation, there is a common term, $-v_0/15$, on both sides that may be canceled. What remains may be rearranged to yield

$$\frac{v_1 - 6}{15} + 0.06v_0^{-0.4} = 0.06v_0^{-0.4}(v_0) - 0.1v_0^{0.6} \tag{2.8.4}$$

Now we can give an interesting interpretation to this last equation. A linear circuit that satisfies this same equation is shown in Figure 2.7.

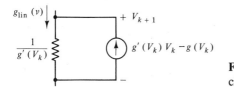

Figure 2.7 Linear model of a nonlinear circuit element.

Write Kirchhoff's current law for this network, and the resultant expression is identical with the one above. In other words, for this particular example, using Newton's method is equivalent to doing a repetitive solution of a linear circuit problem.

The circuit interpretation in the preceding section is a fairly general concept. Assume that we are in the process of doing a Newton-Raphson solution, and we are at the point where our latest solution estimate is v_k. Then, if we generate a linear approximation to $g(v)$ (the function that describes our nonlinear element) around the point $v = v(k)$, we will have

$$g_{\text{lin}}(v) = g(v_k) + g'(v_k)(v - v_k) \tag{2.8.5}$$

Since the units of $g(v)$ are amperes (current), we can give a "current" interpretation to Eq. (2.8.5). In Figure 2.8, if we sum the currents through the two elements and equate that sum to $g_{\text{lin}}(v)$, we obtain Eq. (2.8.5).

Now, we can examine what happens when we apply Newton's method to our example circuit. We have

$$f(v) = \frac{v - 6}{15} + 0.1v^{0.6} = 0 \tag{2.8.6}$$

and the iteration formula is

$$v_{k+1} = v_k - \frac{f(v_k)}{f'(v_k)} \tag{2.8.7}$$

However, we know the derivative of $f(v)$

$$f'(v) = \frac{1}{R} + g'(v) \tag{2.8.8}$$

So we have

$$v_{k+1} = v_k - \frac{f(v_k)}{(1/R) + g'(v_k)} \tag{2.8.9}$$

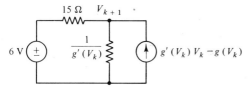

Figure 2.8 Circuit incorporating a linear model of a nonlinear element.

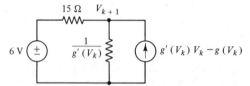

Figure 2.9 Equivalent linear circuit obtained from Newton-Raphson formula.

If we multiply through by the term $(1/R) + g'(v)$ and rearrange, we can obtain

$$(v_{k+1} - v_k)\left[\frac{1}{R} + g'(v_k)\right] = -f(v_k) = \frac{-(v_k - 6)}{R} - gv_k \qquad (2.8.10)$$

Rearranging,

$$\frac{v_{k+1} - 6}{R} + (v_{k+1} - v_k)g'(v_k) = -gv_k \qquad (2.8.11)$$

This last expression can be given a circuit interpretation. If we write Kirchhoff's current law for the circuit in Figure 2.9, we obtain the result found in Equation (2.8.11). Moreover, the components that come from the nonlinear element are the same components that come about when we do a linear expansion around the last solution estimate, v_k.

Let's consider this. What we have is a linear expansion about the most recent guess. This linear expansion generates a linear circuit model that can be used to compute the next step in a Newton-Raphson process. By solving the linear circuit with the nonlinear element replaced by its "equivalent" linear model, we can do one step in the Newton-Raphson process. By doing a number of linear circuit analyses, we can do a number of steps in the Newton-Raphson process. We only need to be able to generate the components in the linear model. The conceptual stages in generating a model are as follows:

1. Obtain a mathematical model (nonlinear?) for the device.
2. Express current (voltage) as a function of voltage (current) using a Taylor series expansion, retaining only first-power (linear) term.
3. Identify terms with a Norton (Thévenin) equivalent circuit model.

As an example of this linear model, let us go back to the semiconductor diode. We have used the nonlinear mathematical model

$$i = g(v)$$
$$= I_0(e^{mv} - 1) \qquad (2.8.12)$$

where $m = q/kT$. If we do a linear expansion around an operating point, v_k, we have

$$i_{k+1} = g(v_{k+1})$$
$$= g(v_k) + (v_{k+1} - v_k)g'(v_k) \qquad (2.8.13a)$$
$$= I_0(e^{mv_k} - 1) + (v_{k+1} - v_k)mI_0e^{mv_k} \qquad (2.8.13b)$$
$$= I_0(e^{mv_k} - 1) - mv_ke^{mv_k} + v_{k+1}mI_0e^{mv_k} \qquad (2.8.13c)$$

So in our linear model we have

$$\text{Current source} = I_0(e^{mv_k} - 1) - mv_k e^{mv_k} \qquad (2.8.14a)$$
$$\text{Conductance} = mI_o e^{mv_k} \qquad (2.8.14b)$$

PROBLEM 2.12

Here are two devices for you to find linear equivalent models for use in the Newton-Raphson algorithm. (They are oldies, but goodies.)

1. A light bulb has a resistance that increases as current and voltage increase because the temperature of the filament increases and resistance is an increasing function of temperature. Mathematically, we might have something like

$$i = 0.082 v^{0.6} \text{ A}$$

Find the linear model

2. A vacuum tube diode has the voltage-current expression

$$i = 0.01 v^{3/2} \text{ A}$$

Find the linear model.

2.9 NEWTON-RAPHSON TECHNIQUES FOR MORE GENERAL CIRCUITS

Once we have come to grips with a circuit interpretation of Newton's method for the simple circuit we have been using as an example, we are ready to look at more general circuits. We do that by looking first at how we must proceed to generalize Newton's method.

Let us consider a slightly more complex circuit than the one we have been considering. Figure 2.10 gives a circuit that is equivalent to the previous example but has two nodes. The additional node comes about because we replaced the 6-V source and the 15-Ω resistor by a more complex, but equivalent, circuit. However, we now have two nodes—and that complicates matters.

If we write Kirchhoff's current law, we have two equations:

$$\frac{v_1 - 12}{20} + \frac{v_1}{20} + \frac{v_1 - v_2}{5} = 0 \qquad (2.9.1a)$$

$$\frac{v_2 - v_1}{5} + g(v_2) = 0 \qquad (2.9.1b)$$

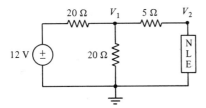

Figure 2.10 A multinode nonlinear circuit.

This set of equations certainly looks even worse than what we considered earlier. With two simultaneous equations (one of which is nonlinear) to solve, the complexity is growing.

These two equations are a specific example of most common situations. In general, we would have as many equations as we have nodes, and each equation could be nonlinear. If we have N nodes, we would have

$$f_1(v_1, v_2, \ldots, v_N) = 0$$
$$f_2(v_1, v_2, \ldots, v_N) = 0$$
$$\vdots$$
$$f_N(v_1, v_2, \ldots, v_N) = 0 \qquad (2.9.2)$$

This set of simultaneous nonlinear equations can be expressed in vector form as

$$\mathbf{F(V)} = \mathbf{0} \qquad (2.9.3)$$

where

$$\mathbf{V} = (v_1, v_2, \ldots, v_N)^{\mathrm{T}} \qquad (2.9.4)$$

and

$$\mathbf{F} = (f_1, f_2, \ldots, f_N)^{\mathrm{T}} \qquad (2.9.5)$$

Thus $\mathbf{F(V)}$ is a vector-valued function of the vector of node voltages $\mathbf{V}$.

In this multidimensional case, we can proceed in a manner reminiscent of the development used in the single-dimensional case. That is,

1. Do a linear expansion of the function $\mathbf{F(V)}$ around the most recent estimate of the root, $\mathbf{V}_k$.
2. Solve for a better root estimate, $\mathbf{V}_{k+1}$, using the linear expansion from 1.
3. Repeat steps 1 and 2 as often as necessary.

First, we do a linear expansion around the most recent root estimate, $\mathbf{V}_k$. Start by considering each component function of $\mathbf{F(V)}$ separately.

$$f_1(v_1, v_2, \ldots, v_N) = f_1(v_{1_0}, v_{2_0}, \ldots, v_{N_0})$$
$$+ \frac{\partial f_1}{\partial v_1}(v_1 - v_{1_0}) + \frac{\partial f_1}{\partial v_2}(v_2 - v_{2_0}) + \cdots \frac{\partial f_1}{\partial v_N}(v_N - v_{N_0}) \qquad (2.9.6)$$

There are similar expressions for the other functions in the vector of functions, $\mathbf{F}$.

It becomes clear that we will not be able to see things as well if we don't soon adopt a simpler notation. We can see things more clearly by expressing all this in a vector matrix notation.

The linear behavior of all N functions near the point $\mathbf{V}_0$, can be summed up as follows:

$$\mathbf{F}_{lin}(\mathbf{V}) = \mathbf{F}(\mathbf{V}_0) + J(\mathbf{V} - \mathbf{V}_0) \tag{2.9.7}$$

where

$$J = \text{Matrix of partial derivatives (Jacobian matrix)} \tag{2.9.8}$$

Now, we use the linear expansion to estimate where the function, $\mathbf{F}(\mathbf{V})$, is zero by solving for where the linear approximation is zero. In an interative scheme, we would assume that we have generated a kth estimate, $\mathbf{V}_k$, and we use the linear approximation to solve for the $(k + 1)$th estimate.

$$\mathbf{F}_{lin}(\mathbf{V}_{k+1}) = \mathbf{F}(\mathbf{V}_k) + J(\mathbf{V}_k)(\mathbf{V}_{k+1} - \mathbf{V}_k) \tag{2.9.9}$$

Solving for the point at which $\mathbf{F}_{lin}(\mathbf{V}_{k+1}) = 0$, we find

$$\mathbf{V}_{k+1} = \mathbf{V}_k - J(\mathbf{V}_k)^{-1}\mathbf{F}(\mathbf{V}_k) \tag{2.9.10}$$

This last expression is the multidimensional version of Newton's method.

In the general form just discussed, the Newton-Raphson algorithm depends on being able to formulate the functions, $f_1, \ldots, f_N$ and derivatives of those functions at every node. That formulation may be a difficult computational problem, which we can circumvent by using a linear model for all nonlinear elements and then doing a repetitive linear solution, changing the linear model as each new iterative point is computed. In Appendix D we find a program segment that adds the linear diode model to the circuit array descriptions, and then repetitively solves the resultant linear circuit.

The nonlinear solution subroutine NONLINSOLVE makes several assumptions. First, it uses a standard form for a diode, using $m = 40$ and $I_0 = 10^{-9}$. Those might not be the parameter values you want, and there should be provision for changing them. More important, however, NONLINSOLVE assumes that the diode voltage will never go above 0.6 V, so it assumes that voltage value across the diode whenever the computed value exceeds 0.6 V. That prevents the slow approach to the solution that we found earlier. Also, NONLINSOLVE never uses a value less than -1 for the voltage across the diode. That is enough to "turn off" the diode and prevents overflow in the exponential function, but it still permits accurate computation (in one iteration) of the back voltage across a back-biased diode.

PROBLEM 2.13

Calculate analytic expressions for the linear model of the circuit in Figure 2.10 and program your results to perform the Newton-Raphson iterations.

PROBLEM 2.14

Determine how the computed value of voltage across the diode depends upon the value of m. Do your computations for the circuit shown, using $R = 100\ \Omega$ and $R = 10\ k\Omega$.

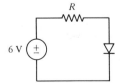

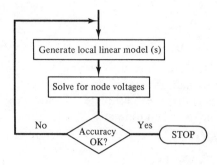

Figure 2.11 Flowchart for multinode nonlinear (Newton) solution.

In the general form just discussed, the Newton-Raphson algorithm can be used only when we are able to formulate the functions $f_1, \ldots, f_N$ and their derivatives at every node. That formulation is a difficult computational problem. We can circumvent that problem by using a linear model for all nonlinear elements and then doing a repetitive solution, changing the linear model as each new iterative point is computed.

A flowchart showing the process of using the Newton algorithm in a circuit with many nonlinearities is given in Figure 2.11. The only item on that chart that is new is generation of the linear model for the nonlinear element. The rest of the numerical techniques for solution of the resulting linear circuit were discussed previously in Chapter 1.

Appendix D gives the subroutine NONLINSOLVE, which does this sort of nonlinear analysis repetitively, implementing the flowchart of Figure 2.11 inside a loop.

The algorithm defined by Figure 2.11 is really the heart of the method used for much more complex networks. Various forms of transistors can be embedded in the networks we want to analyze. In all cases, the algorithm for numerical solution will be to take a localized linear model, linearized around the most recent solution, and use it iteratively to compute a new solution—one that converges to the node voltages.

2.10 THE SECANT METHOD

There are certain problems that are almost unworkable using Newton's method. A classic example is trying to find the zero of the arctangent function. Figure 2.12 shows a graph of the sort of function that can cause trouble. For values of

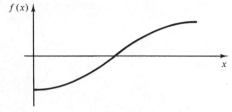

Figure 2.12 A troublesome function.

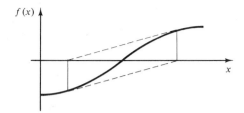

Figure 2.13 Newton algorithm applied to a troublesome function.

x far from the zero the derivative, $f'(x)$, is small—approaching zero. However, a small derivative will produce a large change in the location of x if we use Newton's algorithm. If that large change takes us far beyond the zero to a point where the derivative is also small, then we have the possibility of getting a diverging iteration, as shown in Figure 2.13.

We can get a numerical solution to this sort of problem using a method related to the Newton algorithm, the secant method. There are several versions of this method. Here, we will assume that we have one zero crossing, and that we have measured two points on the function, $f(x_l)$, and $f(x_r)$, where x_l is the leftmost of the two points and x_r is the rightmost. Now construct a straight line (secant) that passes through the two measured points, and then solve for the point at which the straight line passes through zero. We can use x_0 to decrease our search interval, (x_l, x_r). If we have a decreasing function, as shown in Figure 2.14, then we can eliminate either the left-hand portion or the right-hand portion of the interval (x_l, x_r) depending upon whether $f(x_0)$ is positive or negative.

If we repeat this procedure a number of times, then we should be able to converge to the root of the function. This iterative procedure will eventually get close enough to the zero crossing to be acceptable. Although this convergence might take longer than the straightforward Newton-Raphson algorithm, it may converge in instances where the Newton algorithm fails to converge. (Actually, the convergence is slower, but still quadratic. See Section 2.6 of Dennis and Schnabel [8] for a discussion of the convergence properties of secant methods.)

As an example of a circuit problem using the secant method, consider adjusting the resistor R_2 in the circuit of Figure 2.15 to obtain 4 volts across the output terminals.

In this circuit, as R_2 increases, the output voltage, V_{out}, decreases, and the function $V_{out}(R_2)$ turns out to be concave. If we want to compute the resistance

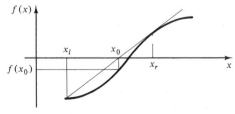

Figure 2.14 One step of the secant method.

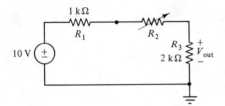

Figure 2.15 An example network.

value that gives some particular value of output voltage, V_{desired}, then we can form a function:

$$f(R_2) = V_{\text{out}}(R_2) - V_{\text{desired}}$$

This function is zero when V_{out} equals V_{desired}. Figure 2.16 shows the function and several iterates of the secant method. For this kind of concave function, the leftmost point of the search interval never gets changed.

Dennis and Schnabel [8] discuss rate of convergence of the secant method, and compare the secant method's rate with that of the Newton-Raphson algorithm. Typically, we expect the secant method to be somewhat slower, but still faster than a search or a functional iteration method.

PROBLEM 2.15

Consider using the secant method to find a point on a resonance curve. Assume that you have the frequency response function:

$$G(f) = \frac{f_n^2}{-f^2 + j2\zeta f_n f + f_n^2}$$

where

$f_n = 1000$ Hz (natural frequency),
$\zeta = 0.05$

Using the secant method, determine where

(a) $|G(f)| = 2.0$
(b) $|G(f)| = 0.2$

2.11 THE EBERS-MOLL TRANSISTOR MODEL

We have now covered enough material that we can model binary junction transistors (BJTs) using the elements we have modeled so far. BJTs are important

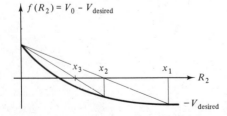

Figure 2.16 Applying the algorithm to the example network of Figure 2.15.

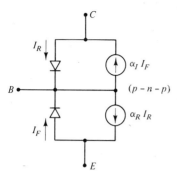

Figure 2.17   The Ebers-Moll transistor model.

components even though integrated circuits now command the most important place in electronics. However, BJTs are still common circuit elements, and they also appear in models of integrated circuits—for example, operational amplifiers [10].

The simplest model for a BJT is a linear model using controlled sources and resistors. However, the linear model neglects many important nonlinear effects that cause saturation, for example. The simplest, and probably most widely used, model of a BJT is the Ebers-Moll model [11, 12]. Figure 2.17 shows it in basic form.

The Ebers-Moll model has two current-controlled current sources (CCCSs)—to account for the transistor-amplifying action of the device—and two diodes—to account for the semiconductor junction behavior of the base-emitter and base-collector junctions. With this basic model it is possible to account for much of the DC and low-frequency nonlinear behavior of a BJT. However, it neglects any charge storage in the junctions, for example, and will not be valid for short-time or high-frequency predictions or calculations.

To use the Ebers-Moll model in a circuit analysis program does not present any great difficulty. If our program keeps track of transistor connections, then each and every transistor can be replaced by two diodes and two CCCSs with appropriate connections. The only complicating factor is the use of four two-terminal components to model a single three-terminal component.

If we need a more complete or accurate description of a transistor, there do exist models that account for further details of transistor behavior. The Gummel-Poon model [13, 14] accounts for frequency dependent behavior due particularly to charge storage effects in the junctions. The default model in SPICE is the Gummel-Poon model, and it is widely used elsewhere.

2.12 AN ODD WAY THINGS CAN GO WRONG

Electrical engineers have extensive experience with various systems, and they tend to project their experience into the numerical techniques they use. As a result, when things go wrong we tend to expect exponential growth in an error, or an oscillation. That may not always be what happens when an algorithm self-

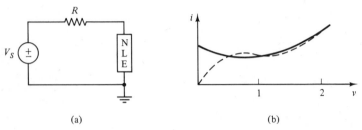

Figure 2.18 An "attractive" nonlinear circuit and circuit element.

destructs. In this section we will examine an artificial circuit, designed to show the numerical problems, and we will let the interested reader pursue the topic in one of the miniprojects.

The circuit we will examine is the same one we started the chapter with. Here the nonlinear element is chosen to have a peculiar *V-I* characteristic; that is,

$$I = f(V) = \frac{V_s - 4LV(1 - V)}{R} \qquad (2.12.1)$$

The circuit and the nonlinear characteristic are shown in Figure 2.18. This particular nonlinear characteristic is completely fictitious, but the interesting portion of the nonlinearity closely resembles that of a tunnel diode. [15]. (A tunnel diode would look more like the nonlinearity shown as a broken line in the figure.)

An attempt was made to calculate the output voltage using the functional iteration method. Results of this abortive attempt are random. (Or at least they look that way to a casual observer. This numerical process is deterministic nonetheless.) As the depth of the valley in the nonlinearity increases (the valley bottom gets lower) the numerical process changes from convergence to oscillation to noise generation. The shift from convergence to oscillation is abrupt, as is the shift from oscillation to noise generation. In the region where oscillation is observed, as the depth of the valley increases, the period lengthens from two calculations to four to eight and so on.

A reading of Hofstadter's *Scientific American* article [6] sheds some light on these results. The difference equation being solved is one that has a "strange attractor" of the sort Hofstadter describes, and the particular nonlinearity used in this example is chosen to give Hofstadter's example exactly.

Although this example has been carefully crafted to exhibit bizarre behavior, it doesn't seem so bad that we could absolutely rule out the possibility of it ever occurring in a realistic problem. The nonlinearity is not that complex, and we could easily construct this kind of example in SPICE. (Although SPICE does not use functional iteration, it does use the Newton-Raphson algorithm.) As we leave this subject, we simply need to wave a few warning flags, and invite the interested student to look further by working out Miniproject 5.

REFERENCES

1. R. Tomovic, *Introduction to Nonlinear Automatic Control Systems,* Wiley, New York, 1966. (Tomovic discusses many of the phenomena that can occur in nonlinear networks including, particularly, dynamic phenomena.)
2. A. W. Willson, Jr., Some aspects of the theory of nonlinear networks, *Proceedings of the IEEE,* vol. 61, pp. 1092–1113, August 1973.
3. R. J. Duffin, Nonlinear networks IIa, *Bulletin of the American Mathematical Society,* vol. 53, pp. 963–971, October 1947.
4. C. A. Desoer and J. Katzenelson, Nonlinear RLC networks, *Bell System Technical Journal,* vol. 44, pp. 161–198, January 1965.

   References [2]–[4] are reprinted in *Nonlinear Networks: Theory and Analysis,* edited by A. N. Willson, Jr., published by IEEE Press. There are numerous other good articles in this collection, but these three are eminently readable. The interested reader may wish to peruse the rest of the collection.

5. L. O. Chua and P-M Lin, *Computer Aided Analysis of Electronic Circuits: Algorithms and Computational Techniques,* Prentice-Hall, Englewood Cliffs, NJ, 1975, pp. 46–50.
6. D. Hofstadter, Metamagical themas, *Scientific American,* November 1981, pp. 22–43.
7. Richard W. Daniels, *An Introduction to Numerical Methods and Optimization Techniques,* North-Holland, New York, 1978.
8. J. E. Dennis, Jr., and R. B. Schnabel, *Numerical Methods for Unconstrained Optimization and Nonlinear Equations,* Prentice-Hall, Englewood Cliffs, NJ, 1983.
9. P. E. Gill, W. Murray, and M. H. Wright, *Practical Optimization,* Academic Press, New York, 1981.
10. G. J. Herskovitz and R. B. Schilling (Eds.), *Semiconductor Device Modeling for Computer-Aided Design,* McGraw-Hill, New York, 1972.
11. G. R. Boyle, B. M. Cohn, D. O. Pederson, and J. E. Solomon, Macromodeling of integrated circuit operational amplifiers, *IEEE Journal of Solid-State Circuits,* vol. SC-9, December 1974, pp. 353–364.
12. J. J. Ebers and J. L. Moll, Large-signal behavior of junction transistors, *Proceedings of the IRE,* vol. 42, December 1954, pp. 1761–1772. (This paper could be considered the definitive paper on junction transistor models. However, the model itself is found in many electronics texts. See [13], for example.)
13. A. S. Sedra and K. C. Smith, *Microelectronic Circuits,* Holt, Rinehart and Winston, New York, 1982.
14. H. K. Gummel and H. C. Poon, An integral charge control model of bipolar transistors, *Bell System Technical Journal,* vol. 49, May–June 1970, pp. 827–851.
15. D. J. Comer, *Modern Electronic Circuit Design,* Addison-Wesley, Reading, MA, 1978, pp. 46–48.

MINIPROJECTS

1. Sometimes it is desirable to model a two-terminal device with a polynomial relationship between voltage and current. Say you have $I = A + BV + CV^2$. Devise and program a model for use in the Newton algorithm for this polynomial relationship.

2. Many microcomputer systems suffer from lack of full mathematical coprocessing ability, and cannot compute transcendental, trigonometric, or power functions. The Newton algorithm offers a way to compute square roots using only the operations of addition, subtraction, multiplication, and division. These are operations available on many present-day chips.

   (a) Derive a square root algorithm using the Newton method to solve the equation:

   $$x^n = C \qquad \text{or} \qquad f(x) = x^n - C = 0$$

   (b) Derive an algorithm to take the "0.4 power" of a number. Can this algorithm be implemented using only addition, subtraction, multiplication, and division?

   (c) Program the square root algorithm in FORTRAN and check convergence properties of the algorithm.

3. Consider the frequency response function

   $$G(j\omega) = \frac{1}{((1 + j\omega)(1 + 2j\omega)(1 + 5j\omega))}$$

   (a) Using the secant method, determine the ω for which this function becomes $-180°$. Repeat for $-150°$.

   (b) Generalize your program to N real poles.

   (c) Generalize your program to N real poles and M real zeros.

4. In control system analysis, it is frequently useful to be able to locate the frequency at which a frequency response function has unity gain (0 dB).

   (a) Start with the "classic" second-order system

   $$G(s) = \frac{Ab}{s^2 + as + b}$$

   Write a program that uses Newton's algorithm to find when $|G(j\omega)| = 1$. Compare the secant method applied to the same problem.

   (b) Use decibels (dB) instead of $|G|$ in the algorithm, and compare your results with those of part (a).

5. In Section 2.12 we encountered a nonlinearity that produced random results using the functional iteration scheme.

   (a) Is it possible to construct a circuit that will exhibit the same behavior when using Newton-Raphson?

   (b) If it is possible, then construct the model and test Newton-Raphson on it (and do it in SPICE, if SPICE is available where you are working).

Chapter 3

Simulation I

3.1 COMPUTATION OF TRANSIENT RESPONSE

How a system responds to inputs is the most important item of information we can have about a circuit or system. We can compute the response of a circuit for a number of different kinds of inputs. Usually we employ steps, ramps, and sinusoids. For any given type of input there are usually different parameters that can be defined and measured. For steps, we can measure the rise time of the step response, overshoot, and the like. For sinusoidal inputs, we can plot the amplitude and phase of the response. Lore and mythology based on frequency response parameters are used when purchasing diverse systems such as stereo amplifiers and pneumatic control systems. All of these characteristics of time responses are important. In many designs it is important to be able to predict different facets of the time response of a system.

In this chapter we want to consider some ways of calculating the time response of a circuit. We will focus first on simple methods applied to simple systems, with an eye toward later generalizing our methods to more complex systems and methods. There are not only different methods of calculating transient response, but there are many different ways to implement these methods. As we proceed, we will first look at stand-alone programs that implement particular algorithms. Later we will investigate embedding an algorithm in a circuit analysis program.

Before we can begin calculating response, we need to realize that we need two items of information before we can begin. First, we have to have some sort of mathematical description of our system, a description of the system that lends itself to some sort of response calculation. Second, we need some sort of numerical method or algorithm for calculation of the system's response. We hope that our method for calculating response and our system description will work together.

3.2 A SIMPLE ALGORITHM AND A SIMPLE CIRCUIT

To get started on response computation, let's take a simple resistor-capacitor network, and examine the problem of computing the time response of this circuit for an arbitrary input. The circuit we are using, which is shown in Figure 3.1, has the following differential equation relating its input and output:

$$RC\dot{V}_o(t) + V_o(t) = V_{in}(t)$$

We will assume that we know the input voltage, $V_{in}(t)$, and that we are trying to compute the output voltage, $V_o(t)$. Let's be precise about what we expect to get. After all, doing things digitally, we don't really expect to get a formula for the output voltage. Rather, what we expect is that we will be able to compute points of the function, $V_o(t)$, for some discrete set of time values. Usually, we would expect these time values (sample times) to be uniformly spaced. That is what we will assume we are doing.

We will also have to assume that we have the initial value of the output voltage, $V_o(0)$. Now, imagine that we are starting at that initial value, and trying to compute the value of the output voltage, $V_o(t)$, some time, DT, later. Then we can compute an approximate value for the output voltage at time DT as

$$V_o(DT) = V_o(0) + DT\dot{V}_o(0) \tag{3.2.1}$$

What we are doing is a linear extrapolation. We are taking just the first terms in a Taylor series expansion for $V_o(DT)$, neglecting terms higher than those beyond the first derivative. We recognize that we could get better accuracy by taking the second-, or even third-derivative terms, but if the time interval DT is small, we may be able to get acceptable accuracy with just the linear term.

This method of approximately calculating the value of $V_o(DT)$, though relatively simple, does go by the name of "Euler integration algorithm." The Euler integration algorithm can be used for other first-order differential equations in which the derivative can be isolated and solved for. It can also be generalized to higher-order systems. Finally, in the next chapter, we will consider more exact and complex integration algorithms—including, for example, Runge-Kutta methods.

Now, from the differential equation relating input and output, we can get an expression for the derivative of the output voltage:

$$\dot{V}_o(t) = \frac{V_{in}(t) - V_o(t)}{RC} \tag{3.2.2}$$

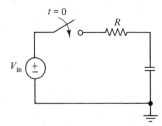

Figure 3.1   A simple circuit with one state.

Putting this expression into our approximate expression for the output, we have

$$V_o(DT) = \left(1 - \frac{DT}{RC}\right)V_o(0) + \frac{V_{in}(0)DT}{RC} \qquad (3.2.3)$$

Let's try this with some numerical values. Assume we have

$$RC = 0.1 \text{ s} \qquad (3.2.4a)$$
$$DT = 0.01 \text{ s} \qquad (3.2.4b)$$
$$V_o(0) = 0.0 \text{ V} \qquad (3.2.4c)$$
$$V_{in}(t) = 10.0 \text{ V} \qquad \text{(constant)} \qquad (3.2.4d)$$

Now, if we compute the response using these values, we get the results shown in Table 3.1.

We can note two important points about the results in Table 3.1. First, these results are tending to approach the correct steady-state value—that is, 10. Second, the intermediate values look "reasonably" accurate. For example, at one time constant (0.1 s) the correct value is $6.321v$, and our computation gives us $6.513v$.

The data from Table 3.1 is also plotted in Figure 3.2, using a line-printer plotter routine.

We can look at what happens if we use a different integration interval. Table 3.2 gives results for a DT of 0.02 s, twice as long as the 0.01 s used in Table 3.1. The same data is again plotted as Figure 3.3.

TABLE 3.1

TIME	STATES
.000000E+00	.000000E+00
.100000E−01	.100000E+01
.200000E−01	.190000E+01
.300000E−01	.271000E+01
.400000E−01	.343900E+01
.500000E−01	.409510E+01
.600000E−01	.468559E+01
.700000E−01	.521703E+01
.800000E−01	.569533E+01
.900000E−01	.612579E+01
.100000E+00	.651322E+01
.110000E+00	.686189E+01
.120000E+00	.717570E+01
.130000E+00	.745813E+01
.140000E+00	.771232E+01
.150000E+00	.794109E+01
.160000E+00	.814698E+01
.170000E+00	.833228E+01
.180000E+00	.849905E+01
.190000E+00	.864915E+01
.200000E+00	.878423E+01

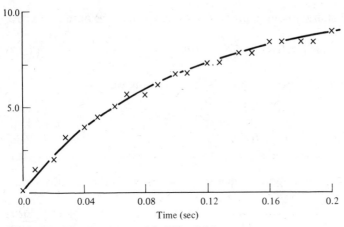

Figure 3.2 Time response with $DT = 0.01$ s.

Here we see the same tendency to approach the right steady state, but the intermediate computations are not as accurate. Compare any value and the results for $DT = 0.02$ are less accurate than the results obtained using $DT = 0.01$.

Let us continue using different values for the integration interval, DT. Tables 3.3, 3.4, and 3.5 give results using $DT = 0.1$, $DT = 0.2$, and $DT = 0.5$, respectively. For $DT = 0.1$ we find that the computed result jumps directly to the steady-state value, as shown in Figure 3.4. For $DT = 0.2$ we find an oscillation that doesn't die out, as shown in Figure 3.5. For $DT = 0.5$ we find increasing oscillations. Clearly, there is a limit beyond which we get worse than just inaccurate computations; we get divergence away from the correct solution.

In the next section we will look at this phenomenon of *numerical instability* in a little more detail, as well as worrying about the general problem of accuracy in this kind of computation.

TABLE 3.2

TIME	STATES
.000000E+00	.000000E+00
.200000E-01	.200000E+01
.400000E-01	.360000E+01
.600000E-01	.488000E+01
.800000E-01	.590400E+01
.100000E+00	.672320E+01
.120000E+00	.737856E+01
.140000E+00	.790285E+01
.160000E+00	.832228E+01
.180000E+00	.865782E+01
.200000E+00	.892626E+01

$DT = 0.02$

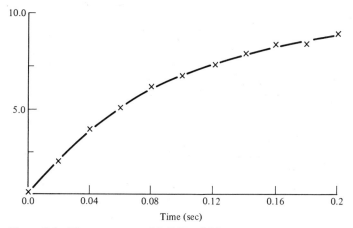

Figure 3.3 Time response with $DT = 0.02$ s.

We should note that we can get better accuracy if we use a smaller integration interval, DT. However, if we use a small value of DT, then we will have to do more computations to cover any particular total interval. For example, if we want to compute the results from zero to one-half second, then if we use 0.01 s for DT, we need to do 50 iterations of the Euler algorithm. If we wanted more accuracy, then we would choose a smaller DT, say 0.001 second, and in this case we would do 500 iterations. More accuracy is achieved only by doing more computations.

However, if we carry this thought further, we will find that eventually we are doing so many computations that we don't want to see them all. Then we would have to nest loops in our control program so that we print the results only every so often. Even worse, there is a limit to what we can achieve. Sooner or later, we will begin getting inaccuracy in our additions in the integration algorithm. When the increment added to the state becomes small enough, we

TABLE 3.3

TIME	STATES
.000000E+00	.000000E+00
.100000E+00	.100000E+02
.200000E+00	.100000E+02
.300000E+00	.100000E+02
.400000E+00	.100000E+02
.500000E+00	.100000E+02
.600000E+00	.100000E+02
.700000E+00	.100000E+02
.800000E+00	.100000E+02
.900000E+00	.100000E+02
.100000E+01	.100000E+02

$DT = 0.1$

TABLE 3.4

TIME	STATES
.000000E+00	.000000E+00
.200000E+00	.200000E+02
.400000E+00	.160933E−05
.600000E+00	.200000E+02
.800000E+00	.351667E−05
.100000E+01	.200000E+02
.120000E+01	.351667E−05
.140000E+01	.200000E+02
.160000E+01	.351667E−05
.180000E+01	.200000E+02
.200000E+01	.351667E−05

DT = 0.2

will encounter roundoff problems, and then further reduction of the integration interval will begin to cause increased inaccuracy. Somewhere in the middle is an optimum point. Usually, the optimum is broad enough that there is a wide range of acceptable integration intervals, but problems are possible.

Most important to remember is that CPU time is never free, and sooner or later we will reach a level where the cost of increased computation time becomes intolerable. The cost of increased computation is not always going to be balanced by the benefits of the resultant increase in accuracy whenever we cut the integration interval, and the choice of integration interval is usually a compromise decision.

ALGORITHM

The Euler integration algorithm can be summarized as follows:

1. Compute

$$x(t + DT) = x(t) + \dot{x}(t) \cdot DT$$

and increment t to $t + DT$.

TABLE 3.5

TIME	STATES
.000000E+00	.000000E+00
.500000E+00	.500000E+02
.100000E+01	−.150000E+03
.150000E+01	.650000E+03
.200000E+01	−.255000E+04
.250000E+01	.102500E+05
.300000E+01	−.409500E+05
.350000E+01	.163850E+06
.400000E+01	−.655350E+06
.450000E+01	.262145E+07
.500000E+01	−.104858E+08

DT = 0.5

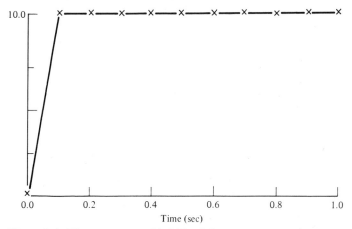

Figure 3.4 Time response with $DT = 0.1$ s.

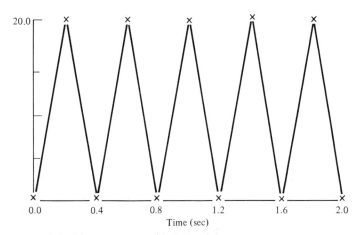

Figure 3.5 Time response with $DT = 0.2$ s.

2. Repeat step 1 until $t \geq T_{stop}$.

The iteration is carried out until some predetermined time interval has been covered.

PROBLEM 3.1

Using the Euler algorithm compute the response of the inductor-resistor circuit shown. Verify that oscillations develop whenever DT is chosen to be larger than twice the time constant.

$$\frac{di}{dt} = \frac{-R}{L}i + \frac{V_{in}}{L} \qquad \begin{array}{l} L = .001 \text{ H} \\ R = 2000 \ \Omega \end{array}$$

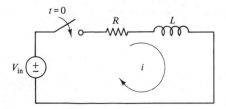

PROBLEM 3.2

A simple model for the growth of a single species is the differential equation:

$$\frac{dP}{dt} = rP$$

This differential equation has a solution that grows exponentially.

The whooping crane population was 51 at one point, and can increase about 4 percent a year (so $r = 0.04$). Using the Euler integration algorithm, determine whether the numerical integration process accentuates or diminishes the instability that is found in this problem, and determine the integration interval that gives either twice or half the theoretical doubling time.

3.3 STABILITY AND ACCURACY IN EULER'S METHOD

Since we can have problems with the accuracy and stability of the Euler algorithm, it would be worthwhile digging a little deeper into these areas. Moreover, the kinds of behavior we find in the Euler algorithm are typical of what we find in other algorithms like Runge-Kutta methods and predictor-corrector methods. The way we will approach the problem of demonstrating the stability problems that might occur will be to do some example calculations for systems that have zero-valued inputs. This simplifies our system to the point that we expose the essential qualities of the dynamics. For linear systems particularly, the stability of the integration algorithm will not depend upon whether or not we have an input.

Let us say that we have a single state equation describing our system. This is written as

$$\dot{x} = \frac{-x}{\tau} \qquad \tau = \text{system time constant} \qquad (3.3.1)$$

This differential equation has no driving term, and the solution for it is

$$x(t) = x(0)e^{-t/\tau} \qquad (3.3.2)$$

If we solve this numerically using Euler's method, we have

$$x(t + DT) = x(t) + DT * \dot{x}(t) \qquad (3.3.3)$$

If we focus on times that are integral multiples of the sample time, DT, and start at $t = 0$, we have

$$x((k+1)\,DT) = x(k\,DT) + DT * x(k\,DT) \qquad (3.3.4a)$$

$$= x(k\,DT) - \frac{DT * x(k\,DT)}{\tau} \qquad (3.3.4b)$$

$$= \left(\frac{1-DT}{\tau}\right) * x(k\,DT) \qquad (3.3.4c)$$

$$= r * x(k\,DT) \qquad (3.3.4d)$$

where

$$r = \frac{1-DT}{\tau} \qquad (3.3.5)$$

Now, imagine this computation starting at the value $x(0)$. Then, we would have

$$x(1) = r * x(0)$$
$$x(2) = r * x(1) = r * r * x(0)$$
$$= r^2 x(0) \qquad (3.3.6)$$

In general, we have

$$x(k) = r^k * x(0) \qquad (3.3.7)$$

If we have $|r| > 1$, then this computed solution will grow without bound, even though the theoretical solution dies out. In order to have $r > 1$, we would have to have

$$T > 2\tau \qquad (3.3.8)$$

However, if $T > 2\tau$, we have instability. For stability, we must have

$$T < 2\tau \qquad (3.3.9)$$

We also need to worry about the accuracy that we will achieve using this method. The upper bound on DT is 2τ, but observing that upper bound just guarantees stability. It does not guarantee that the computation will be accurate. To look at the problem of accuracy, let us focus on the first step of the computation. The actual solution of our differential equation is

$$x(DT) = x(0) * e^{-DT/\tau} \qquad (3.3.10)$$

However, the computed solution is really

$$x(DT) = \left(\frac{1-DT}{\tau}\right) * x(0) \qquad (3.3.11)$$

The factor multiplying the starting value, $x(0)$, in the computed solution should be an exponential. However, what it turns out to be is the first two terms in the series for the exponential. What it should be is

$$e^{-DT/\tau} = 1 - \frac{DT}{\tau} + \frac{(DT)^2}{2\tau^2} + \cdots \qquad (3.3.12)$$

In effect, the Euler integration algorithm drops off the quadratic and higher-degree terms. We could estimate the error in our approximation by using the first term in the series that we drop—that is, the quadratic term, or

$$\frac{(DT)^2}{2\tau^2} \qquad\qquad (3.3.13)$$

This quadratic term should be considerably larger than the next (cubic) term. If we assume that we are doing reasonably accurate computations, any neglected terms should not be too large, and should rapidly decrease. Our first example used a ratio of 0.1 for the interval to time constant ratio, DT/τ. Examining computed values and errors, we find

$$e^{-0.1} = 0.90484 \qquad\qquad (3.3.14a)$$
$$1 - DT/\tau = 0.9000 \qquad\qquad (3.3.14b)$$
$$\text{Est. error} = 0.05 \qquad\qquad (3.3.14c)$$
$$\text{Act. error} = 0.0484 \qquad\qquad (3.3.14d)$$

We need to try to draw some conclusions from this. The first thing we notice is that the error depends upon the square of the integration interval. So, if we need to cut error by a factor of 100, for example, we need only cut the integration interval by a factor of 10. It also seems true that we don't get reasonable accuracy until we get below about one-tenth of the time constant.

Go back to the example computations used in Section 3.2. The time constant in that circuit was 0.1 s ($\tau = 0.1$). We experimentally found sustained oscillations when we used an integration interval DT equal to twice the time constant τ. For larger DT we found numerical instability, and we began to get acceptable accuracy at a value of T equal to one-tenth of the time constant, τ. That specific numerical behavior confirms our general conclusions.

PROBLEM 3.3

The radioactive decay of an atomic species, S, can be modeled with a first-order differential equation:

$$\frac{dS}{dt} = -D * S$$

If D^{-1} is 300 s (5 min) what integration intervals (values of DT) will produce (a) oscillatory solutions, (b) unstable solutions, and (c) reasonably accurate solutions?

EXAMPLE 3.1

Lettuce grows at a rate proportional to the total light received by the plant. A heavier, larger plant intercepts more of the incident radiation and adds weight and leaf area faster than a small plant. Growth is also proportional to the incident light intensity. All of this can be summed up with the growth equation [6]:

$$\frac{dW}{dt} = R(L/L_0)W$$

where R is a rate constant, L is the light intensity referred to a standard intensity L_0, and W is weight.

A program that uses the Euler integration algorithm to solve this differential equation for two different light intensity time variations is as follows:

```
       CHARACTER*1 YESORNO
       WRITE (6,*)' Input the Initial Weight (grams).'
       WRITE (6,*)' (.02 g is typical.)'
       READ (5,*) W
       WRITE (6,*)' How many days until harvesting?'
       READ (5,*) NDAYS
       WRITE (6,*)' Do you want ON/OFF light pattern (Y/N)?'
       READ (5,1000) YESORNO
1000   FORMAT (A1)
       DELTA = 1.69
       R = .69/DELTA
       DT = 1./24.
C
       DO 200 I = 1,NDAYS
          DO 100 NHOUR = 1,24
             FOOTCANDLES = 0.
          IF (YESORNO .EQ. ''Y'') THEN
             IF((NHOUR .GT. 4) .AND. (NHOUR .LT. 20)) THEN
                FOOTCANDLES = 1000.
             ENDIF
          ELSE
             IF((NHOUR .GT. 4) .AND. (NHOUR .LT. 20)) THEN
                FOOTCANDLES = 1000.*SIN((NHOUR-4.)*3.14159/16.)
             ENDIF
          ENDIF
C
          WRATE = R * (FOOTCANDLES/2500.) * W
          W = W + WRATE *DT
100       CONTINUE
200    CONTINUE
       WRITE (6,*)' At harvest, weight = ',W,' grams.'
       CALL EXIT
       END
```

PROBLEM 3.4

Modify the program of Example 3.1 to allow the user to:

(a) Modify the light intensity.
(b) Investigate the effect of pulsed light (on and off) during the 16-hour growing day.

PROBLEM 3.5

Using the program of Example 3.1 as a model, write a program that will permit a user to solve for a whooping crane population in a given year, with the possibility of "poaching" included. You should experimentally determine a reliable value of integration interval, and set that interval automatically, without relying on user specification.

3.4 A PROGRAM STRUCTURE FOR IMPLEMENTATION OF EULER'S METHOD: AN EXAMPLE OF A GENERAL INTEGRATION PACKAGE

Euler's method is just one possible integration algorithm out of many possible. However, even restricting ourselves to that one method doesn't tie everything down completely. The mode of implementation of the algorithm can take many forms. In this section we will look at one way to implement the algorithm. This implementation is commonly employed whenever the user can provide a description of the system dynamics as a collection of state equations. In a later section we will look at another way of implementing the same integration algorithm using a circuit interpretation and a linear circuit analysis program.

The "classical" method of implementing an integration algorithm in a package separates three different functions:

> Data input/output and program control
>
> The integration algorithm
>
> The system description

Usually, the "data input/output and program control" portion (or just "control program") calls the integration algorithm (which is a subroutine). In turn, the integration algorithm subroutine calls the system description subroutine to provide calculated values of state derivatives as necessary.

As an example of the implementation of this concept of separation of function, we will examine the programs used to generate the data we used in Section 3.2. This program uses a subroutine structure to separate the data input function from the control of the simulation. The mainline used to do this calculation is given in Table 3.6.

The different functions performed by the subroutines are as follows:

> SIMIN requests the user to input all the data needed to do the simulation. The user needs to supply an integration interval, DT, and tell the program how many integration steps and how many states there are, and specify initial conditions.
>
> SIMRUN controls the simulation run, printing a header and the simulation results, stopping when finished.
>
> EULER implements the Euler integration algorithm. It gets called by the control program SIMIN, and calls XDOT.

TABLE 3.6

```
DIMENSION X(10)
CALL SIMIN (X,N,DT,NSTEPS)
CALL SIMRUN (X,N,DT,NSTEPS)
CALL EXIT
END
```

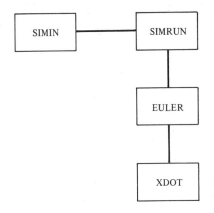

Figure 3.6 Program structure for a simulation program.

XDOT is the subroutine that contains the dynamic description of the system being simulated. XDOT just contains a FORTRAN coding of the state equations.

These FORTRAN modules are given in the next few tables. The overall program structure is shown in Figure 3.6. The first two subroutines (SIMIN and SIMRUN) are shown in Table 3.7.

The routine for Euler integration is shown in Table 3.8, and the implementation of the state equations is done in XDOT, as shown in Table 3.9.

With this kind of modular program structure, we have separated all of the functions. If we want to calculate the response of a different circuit or system, we have to change XDOT, but only XDOT. If we want to change the integration algorithm, we need only substitute the new algorithm in place of EULER. This kind of separation of function is common in a number of packages. In simulation packages like ACSL, CSMP, SL-1, or NDTRAN, the integration algorithm can be independently specified by the user. Conversely, one algorithm can be used with a number of different systems.

With many present operating systems it is possible to build a "command file" that, when executed, will prompt for an XDOT filename, compile that file, and link to the other modules so that anyone can build a simple sort of simulation package.

3.5 USING EULER'S METHOD FOR MORE COMPLEX CIRCUITS

The problems we have done so far have been with just a single-state circuit. However, the most interesting problems involve circuits with many states, not just one state. Indeed, there would not be much need to use digital methods if we were always computing responses of circuits with just one state. Therefore, in this section we will extend our concepts to cover circuits with multiple states.

A simple circuit with two states is given in Figure 3.7. Those states are the

TABLE 3.7

```
C
C*********************************************************************
C
      SUBROUTINE SIMIN (X,N,DT,NSTEPS)
      DIMENSION X(10)
      WRITE(6,*)'' Input Data for the Simulation.''
      WRITE(6,*)'' Input the Number of States.''
      READ (5,*) N
      WRITE(6,*)'' Input the Initial State Values (N States).''
      READ (5,*) (X(J), J=1,N)
      WRITE(6,*)'' Input the Integration Interval (DT).''
      READ (5,*) DT
      WRITE(6,*)'' Input the Number of Steps in the Computation.''
      READ(5,*), NSTEPS
      TIME = DT*NSTEPS
      WRITE(6,*)'' The Total Time in the Run ='', TIME
      RETURN
      END
C
C*********************************************************************
C

      SUBROUTINE SIMRUN (X,N,DT,NSTEPS)
      DIMENSION X(10)
      T = 0.
      WRITE(6,*)'' Time              States''
 1001 FORMAT(1X,8(E12.6,4X))
      WRITE(6,1001), T, (X(I), I=1,N)
      DO 100 IT = 1,NSTEPS
         CALL EULER (X,T,DT,N)
         WRITE(6,1001), T,(X(I), I=1,N)
  100 CONTINUE
      RETURN
      END
C
C*********************************************************************
C
```

TABLE 3.8

```
      SUBROUTINE EULER (X,T,DT,N)
      DIMENSION X(10), XD(10)
C
C  EULER'S METHOD FOR INTEGRATION
C
      CALL XDOT (X,XD,T,N)
      DO 100 I=1,N
         X(I) = X(I)+DT*XD(I)
  100 CONTINUE
      T = T+DT
      RETURN
      END
C
C*********************************************************************
C
```

TABLE 3.9

```
C
C  *******************************
C
       SUBROUTINE XDOT(X,XD,T,N)
       DIMENSION X(10), XD(10)
       XD(1)  =  (10.  -  X(1))/(.1)
       RETURN
       END
C
C  *******************************
C
```

capacitor voltage V_c and the inductor current I_l. If we are going to use the Euler algorithm to compute the response of the circuit, we need to write the equations describing the circuit in state variable format. The state variable format has a set of first-order differential equations, and the Euler algorithm is designed only for first-order equations, since it uses only the first derivative. By writing a set of first-order equations, rather than a higher-order differential equation, we put our mathematical description of the system into a form amenable for use with the Euler algorithm.

For the circuit of Figure 3.7 we can write the state equations without too much difficulty. There is only one loop current, and we can define one state as the loop current since it is the inductor current (and therefore determines how much energy is stored in the inductor). The other state is the capacitor voltage, and it is related to the inductor current as follows:

$$I_l = C\dot{V}_c(t)$$

By writing Kirchhoff's voltage law around the loop, we can also get the other state equation as

$$L\dot{I}_l = V_{in} - RI_l - V_c$$

In this case we are very fortunate in that we are able to write two equations that can be interpreted as a set of state equations. Normally, we wouldn't be this lucky. However, we are going to defer to a later chapter the problem of getting a set of state equations for an arbitrary network, using digital algorithms.

When we have more than one state, we can use the same algorithm for each state. Looking at the EULER subroutine in Table 3.8, we see that the algorithm is implemented just that way.

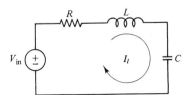

Figure 3.7 A circuit with two states.

TABLE 3.10

```
C
C  ******************************
C
      SUBROUTINE XDOT(X,XD,T,N)
      DIMENSION X(10), XD(10)
C
C  R = 200 OHMS
C  L = .01 HENRIES
C  C = 1 UFARAD
C
      XD(1) =(1.E+6)*X(2)
      XD(2) = 1000. - (20000.*X(2)) - 100.*X(1)
      RETURN
      END
C
C  ******************************
C
```

In any event, let us proceed with the chore of writing a new XDOT sub-routine that will calculate the derivatives we need for this set of state equations. Now, it becomes obvious why we needed a state vector (dimensioned) in our programs. With two states, the state vector components are X_1 and X_2 for V_c and I_l, respectively. The XDOT routine is given in Table 3.10.

PROBLEM 3.6

For the network of Figure 3.7 use the XDOT SUBROUTINE of Table 3.10 to compute the step response of this *RLC* network. Experimentally determine the range of allowable integration intervals, *DT*, for which Euler's algorithm is stable for this network.

PROBLEM 3.7

State equations are given below for the network shown. Since both resistors are 10 kΩ and both capacitors are 0.1 μF, the time constants must be of the order of magnitude of $1/(10 \text{ k}\Omega * 0.1 \text{ }\mu\text{F}) = 10^3$. Experimentally determine the upper limit of *DT* for which stable (albeit oscillatory) behavior is observed in the calculation of the step response using Euler's method.

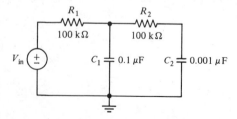

$$\frac{dV_1}{dt} = V_1\left[-\frac{1}{CR_1} - \frac{1}{CR_2}\right] + \frac{V_2}{CR_2} + \frac{V_s}{CR_1}$$

$$\frac{dV_2}{dt} = \frac{V_1}{CR_2} + V_2\left[-\frac{1}{CR_2}\right]$$

3.6 SOME GENERAL STABILITY AND ACCURACY CONSIDERATIONS IN MULTISTATE SYSTEMS

Extending the Euler integration algorithm to systems with multiple states presents a new set of stability problems. Let us consider a simple system with two time constants. The particular transfer function we will examine is

$$G(s) = \frac{2s+1}{s^2+3s+2} = \frac{Y(s)}{U(s)} \tag{3.6.1}$$

There are numerous ways we can represent this particular system. One representation is given in the state equations

$$\frac{dX_1}{dt} = 0*X_1 + 1*X_2 + 2*u \tag{3.6.2a}$$

$$\frac{dX_2}{dt} = -2*X_1 + -3*X_2 + 1*u \tag{3.6.2b}$$

In the standard $d\mathbf{X}/dt = A*\mathbf{X} + \mathbf{B}*\mathbf{U}$ format, we have

$$A = \begin{bmatrix} 0 & 1 \\ -2 & -3 \end{bmatrix} \qquad B = \begin{bmatrix} 2 \\ 1 \end{bmatrix} \tag{3.6.3}$$

and $y = \mathbf{C}^T*\mathbf{X} = X_1$, so

$$\mathbf{C}^T = [1 \quad 0] \tag{3.6.4}$$

If we apply Euler's algorithm to this set of equations it is not immediately clear how the algorithm will behave. However, if we use another representation of the same system and then use the Euler algorithm, we can see clearly how things will go. We can generate another state model for the same system by doing a partial fraction expansion of the transfer function.

$$\frac{2s+1}{s^2+3s+2} = \frac{-1}{s+1} + \frac{3}{s+2} \tag{3.6.5}$$

In diagrammatic form, the two systems in Figure 3.8 are therefore input-output equivalent. A state representation for the "partial fraction" system can then be generated. (Use Q's for the states in this system.)

$$\frac{dQ_1}{dt} = -Q_1 - u \tag{3.6.6a}$$

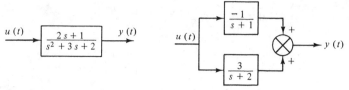

Figure 3.8 Two equivalent systems.

$$\frac{dQ_2}{dt} = -2Q_2 + 3u \qquad (3.6.6b)$$

This partial fraction system representation is interesting because the two states are uncoupled in the sense that each state acts independently.

If we used the Euler integration algorithm on this partial fraction system, each state behaves essentially as an independent system, as follows:

$$Q_1(t + DT) = Q_1(t) + DT[-Q_1(t) - u(t)] \qquad (3.6.7a)$$
$$Q_2(t + DT) = Q_2(t) + DT[-2Q_2(t) + 3u(t)] \qquad (3.6.7b)$$

For a given DT, either of these "subsystems" could be stable or unstable. Clearly, DT must be chosen so that all such subsystem calculations are stable. In this case, DT has to be chosen to be less than the minimum of 2 s and 1 s (twice the time constants) for stability.

Clearly, one question that should be addressed is whether the same behavior occurs in a different system representation. Even though it is not obvious from our discussion thus far, both sets of states are related, and this relationship can be used to answer the questions regarding stability. In particular,

$$\mathbf{X} = \begin{bmatrix} X_1 \\ X_2 \end{bmatrix} = \begin{bmatrix} 1 & 1 \\ -1 & -2 \end{bmatrix} * \begin{bmatrix} Q_1 \\ Q_2 \end{bmatrix} = M * \mathbf{Q} \qquad (3.6.8)$$

Then, since $d\mathbf{X}/dt = A*\mathbf{X} + B*\mathbf{U}$, and $\mathbf{X} = M*\mathbf{Q}$, we have:

$$\frac{d(M*\mathbf{Q})}{dt} = A*M*\mathbf{Q} + B*\mathbf{U} \qquad (3.6.9)$$

Solve for $d\mathbf{Q}/dt$ by premultiplying by M^{-1} to get

$$\frac{d\mathbf{Q}}{dt} = M^{-1}*A*M*\mathbf{Q} + M^{-1}*B*\mathbf{U} \qquad (3.6.10)$$

If we form $M^{-1}*A*M$, we find

$$\begin{bmatrix} 2 & 1 \\ -1 & -1 \end{bmatrix} * \begin{bmatrix} 0 & 1 \\ -2 & -3 \end{bmatrix} * \begin{bmatrix} 1 & 1 \\ -1 & -2 \end{bmatrix} = \begin{bmatrix} -1 & 0 \\ 0 & -2 \end{bmatrix} \qquad (3.6.11)$$

This is the same matrix we obtained in the $\mathbf{Q}$ state equations. A problem left for the interested reader is to verify that $M^{-1}*B$ also gives the control term in those state equations.

These results are best explained using standard linear algebra ideas. The matrix M is a "modal matrix," and the diagonal matrix in the $\mathbf{Q}$-state equations is a diagonal matrix of the eigenvalues of A in the original state equations. The diagonal matrix of eigenvalues is a special case of a Jordan canonical form [1].

We can examine a general case. To take the most understandable approach, let us first consider solving a set of state equations with no input. Assume we have

$$\frac{d\mathbf{X}}{dt} = A * \mathbf{X} \tag{3.6.12}$$

Then, using the Euler integration algorithm, we get

$$\begin{aligned} \mathbf{X}(t+DT) &= \mathbf{X}(t) + DT * A * \mathbf{X}(t) \\ &= (I + A * DT) * \mathbf{X}(t) \end{aligned} \tag{3.6.13}$$

The matrix premultiplying $\mathbf{X}(t)$, $I + A * DT$, has eigenvalues, and can be expressed using a diagonal matrix of eigenvalues if all eigenvalues are distinct.

Let us follow an integration calculation through several steps. Assume the initial state is $\mathbf{X}(0)$, and we compute $\mathbf{X}(DT)$:

$$\mathbf{X}(DT) = (I + A * DT) * \mathbf{X}(0) \tag{3.6.14}$$

and

$$\begin{aligned} \mathbf{X}(2\,DT) &= (I + A * DT) * \mathbf{X}(1) \\ &= (I + A * DT) * (I + A * DT) * \mathbf{X}(0) \\ &= (I + A * DT)^2 * \mathbf{X}(0) \end{aligned} \tag{3.6.15}$$

In general, we would have

$$\mathbf{X}(k\,DT) = (I + A * DT)^k * \mathbf{X}(0) \tag{3.6.16}$$

The behavior of this computed solution is determined by the eigenvalues of the matrix, $(I + A * DT)$. Assume we have the appropriate modal matrix, and express $I + A * DT$ using the modal matrix and a diagonal matrix of those eigenvalues:

$$(I + A * DT) = M * D_{\text{eigen}} * M^{-1} \tag{3.6.17}$$

where M is a *modal* matrix and D_{eigen} is a diagonal matrix of eigenvalues, given by

$$D_{\text{eigen}} = \begin{bmatrix} E_1 & 0 & 0 & \cdots & 0 \\ 0 & E_2 & 0 & & 0 \\ 0 & 0 & E_3 & & 0 \\ \vdots & & & & \vdots \\ 0 & 0 & 0 & & E_n \end{bmatrix} \tag{3.6.18}$$

and $E_k = k$th eigenvalue of A. This is a standard result from matrix theory [2]. If we use this expression for $(I + A * DT)$, then our difference equation solution becomes

$$\begin{aligned} \mathbf{X}(k\,DT) &= (M * D_{\text{eigen}} * M^{-1})^k * X(0) \\ &= M * (D_{\text{eigen}})^k * M^{-1} * X(0) \end{aligned} \tag{3.6.19}$$

Now, since D_{eigen} is a diagonal matrix, then we have the result that any power of D_{eigen} is just a diagonal matrix of powers. In other words,

$$D_{eigen} = \text{Diag}(E_1, E_2, \ldots, E_n) \qquad (3.6.20)$$

and therefore

$$(D_{eigen})^k = \text{Diag}(E_1^k, E_2^k, \ldots, E_n^k) \qquad (3.6.21)$$

If any term in $(D_{eigen})^k$ grows as k increases, then our computation will be numerically unstable. To prevent this we have to have all the eigenvalues, $(E_1, E_1, \ldots, E_n)$, of magnitude less than one—that is, inside the unit circle in the complex plane.

However, it is possible to show that eigenvalues of $(I + A*DT)$ are given by

$$E_i(I + A*DT) = 1 + E_i(A*DT) \qquad i = 1, 2, \ldots, n \qquad (3.6.22)$$

and this is true for any eigenvalue of $(I + A*DT)$, $E_1, E_2, \ldots, E_n$.

For stability, we must have

$$-1 < 1 + E_i(A*DT) < 1 \qquad i = 1, 2, \ldots, n \qquad (3.6.23)$$

Using the left side of this dual inequality, we must have

$$E_i(A*DT) > -2 \qquad i = 1, 2, \ldots, n \qquad (3.6.24)$$

or

$$DT < \frac{-2}{E_i(A)} \qquad i = 1, 2, \ldots, n \qquad (3.6.25)$$

(We assume that all the eigenvalues of $A*DT$ are negative, as they would be in a passive electric circuit, for example.)

This condition is enough to guarantee stability of the computation. It also applies to cases where there are inputs to the circuit (U is nonzero). For good accuracy in the computation, we need to have DT at least an order of magnitude less than the value computed from this inequality, and that must be the case with every eigenvalue.

EXAMPLE 3.2

A system we are simulating has a component with a transfer function of

$$G(s) = \frac{s + 17}{s^3 + 21s^2 + 20s + 0}$$

The poles of this component are at 0, -1, and -20. This indicates that the transfer function of the system is of the form:

$$G(s) = \frac{A}{s} + \frac{B}{s+1} + \frac{C}{s+20}$$

The time constants of the system are 0.05 s, 1 s, and ∞. Consequently, this component in the system would force us to use a DT of 0.005 s or less when we do the simulation.

EXAMPLE 3.3

A system we are simulating has a component with a transfer function of

$$G(s) = \frac{229}{s^2 + 4s + 229}$$

This component has complex poles at $-2 \pm 15j$. Our argument in the preceding section has to be modified with complex poles present. We could look at the magnitude of the poles.

$$|\text{Pole}| = \sqrt{229} = 15.13 \qquad 1/15.13 \approx 0.066$$

From our analysis in this section, this would suggest using a DT of about 0.006 s.

However, we can examine this in another light. With an imaginary part of 15, the component will have a natural response of

$$\sin(15t + \phi)$$

$$15 = \omega_d = 2\pi f_d \qquad T = \frac{1}{f_d} \approx 0.42$$

That response has a period of approximately 0.42 s. That would suggest a DT of about 0.004 s if we wanted to compute at intervals of about one-hundredth of a period. Both of these estimates are in the same "ballpark," and we could probably do reliable computation using either.

PROBLEM 3.8

A circuit with two widely separated time constants (approximately 0.01 s and 0.01 ms) is shown. Table 3.11 gives an XDOT routine for this network. The state equations are as follows:

$$\frac{dV_1}{dt} = \frac{(V_{in} - V_1)/R_1 + (V_2 - V_1)/R_2}{C_1}$$

$$\frac{dV_2}{dt} = \frac{(V_1 - V_2)/R_2}{C_2}$$

and we choose X_1 as V_1, and X_2 as V_2 for the states in our standard representation.

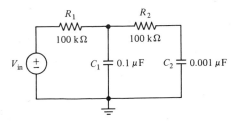

(a) Estimate an integration interval that will give good accuracy for this response computation.

TABLE 3.11

```
C
C  ********************************
C
       SUBROUTINE XDOT (X,XD,T,N)
       DIMENSION X(10), XD(10)
C
       XD(1) = 100.*(10. + X(2) -2.*X(1))
       XD(2) = 100000.*(X(1) - X(2))
       RETURN
       END
C
C  ********************************
C
```

(b) Calculate the number of times XDOT will be called if the total integration time is five times the larger time constant.

3.7 ERROR IN INTEGRATION CALCULATIONS

In doing simulations, at each step we accumulate some roundoff error, and some error due to the approximations made (like including only one term in the Taylor series, and accumulating an error because of the missing terms). After this error occurs, it enters into the next calculation. The next calculation starts from an incorrect value and proceeds to accumulate more error. Thus, each step adds an increment of error, and propagates and changes the errors made in all preceding steps. At any point in the simulation calculation, the total error is the sum of the errors due to all these causes, and as the reader probably suspects, the analysis of the error is complicated by the propagation of the error from step to step.

We can begin to understand the behavior of the "local" error made in a single step of the computation by comparing the expression used in the Euler algorithm with a result available relating to the Taylor series. Provided that all derivatives exist, Reference [3] gives $x(t + DT)$ as

$$x(t+DT) = x(t) + \frac{dx(t)}{dt}\bigg|_{t=t} * DT + \frac{d^2x(t)}{dt^2}\bigg|_{t=T_1} * (DT)^2 \qquad (3.7.1)$$

Here, T_1 is some point in the interval from $x(t)$ to $x(t + DT)$. The only difference between this expression and the Euler integration formula is the second derivative term. Thus, the error in the Euler formula is

$$\frac{d^2x(t)}{dt^2}\bigg|_{t=T_1} (DT)^2 \qquad (3.7.2)$$

The important aspect of the error formula is that it depends upon the square of the integration interval, DT. If the integration interval is halved, it should reduce error to about one-quarter of its previous value.

While the local error is proportional to the square of the integration interval, that really only measures the additional error introduced at the step at time t. What may be more important is whether or not that introduced error grows, decays, or just adds to later local errors. In this section we will examine how error gets added in and then propagates through our calculations. We start by considering a single-time-constant system, and then generalize to more complex linear systems.

Consider applying the Euler algorithm to the first-order system

$$\frac{dx}{dt} = ax$$

Assume we start at time t and are calculating $x(t + DT)$. We have

$$
\begin{aligned}
x_{\text{comp}}(t+DT) &= x(t) + a\,DT\,x(t) \\
&\quad - (1 + a\,DT)x(t) = x(t+DT) - \epsilon(t+DT) \quad (3.7.3)
\end{aligned}
$$

However, the exact expression for $x(t + DT)$ is given by a truncated Taylor series; see Eq. (3.7.1) and Reference [3].

$$
\begin{aligned}
x(t+DT) &= x(t) + a\,DT\,x(t) + \tfrac{1}{2}a^2(DT)^2\ddot{x}(T_1) \\
&= x_{\text{comp}}(t+DT) + \tfrac{1}{2}a^2(DT)^2\ddot{x}(T_1) \quad (3.7.4)
\end{aligned}
$$

Here T_1 is a point between t and $t + DT$. Thus, the error in the computed value is

$$\tfrac{1}{2}a^2(DT)^2\ddot{x}(T_1) \quad (3.7.5)$$

Now the question is to determine what happens to this error as the computation proceeds. Consider carrying the computation one step further in time. We find

$$x_{\text{comp}}(t+2\,DT) = x(t+DT) + a\,DT\,x(t+DT) \quad (3.7.6)$$

However, the true value of $x(t + 2\,DT)$ is given by:

$$x(t+2\,DT) = x(t+DT) + a\,DT\,x(t+DT) + \tfrac{1}{2}a^2(DT)^2\ddot{x}(T_2) \quad (3.7.7)$$

Again, T_2 is a time within the computation interval, $(t + DT, t + 2\,DT)$. Now, we can express this in terms of our computed values and errors.

$$
\begin{aligned}
x(t+2DT) &= (1 + a\,DT)x(t+DT) + \tfrac{1}{2}a^2(DT)^2\ddot{x}(T_2) \\
&= (1 + a\,DT)[x_{\text{comp}}(t+DT) + \tfrac{1}{2}a^2(DT)^2\ddot{x}(T_1)] \\
&\quad + \tfrac{1}{2}a^2(DT)^2\ddot{x}(T_2) \\
&= (1 + a\,DT)^2 x(t) + (1 + a\,DT)[\tfrac{1}{2}a^2(DT)^2\ddot{x}(T_1)] \\
&\quad + \tfrac{1}{2}a^2(DT)^2\ddot{x}(T_2) \quad (3.7.8)
\end{aligned}
$$

The error in $x(t + 2\,DT)$ has two components. One component is the new error introduced in the most recent step. The other component is the error from the first step propagating into the second step.

We can make a few generalizations about the behavior of the propagated error. In this particular case, it seems obvious that the first error will always contribute to the error in every step, and at the nth step we would have

Error contribution of first error at nth step

$$= (1 + a\,DT)^{n-1} \tfrac{1}{2} a^2 (DT)^2 \ddot{x}(T_1) \tag{3.7.9}$$

Other errors would produce a contribution that depended upon the number of computation steps since the error occurred.

If the computation is stable, then the error dies out (as does the computed solution). If the computation is not stable, then the error also grows.

The total error is the sum of the error at any step plus all previous errors, as propagated into that step.

Finally, we should note that we have not really discussed roundoff in this process, so the computation and the system are both linear. If that is true, then our conclusions will still hold if there is an input; that is, the error will decay if the computation is stable.

PROBLEM 3.9

Redo the derivation presented above but add an input to the system so that the system is described by

$$\frac{dx}{dt} = ax + bu$$

Derive an expression that shows the propagated error and new error accumulating.

We must also note that essentially the same derivation of propagated error can be done if the system has more than just one state. In that case we start from the system description

$$\dot{\mathbf{X}} = A\mathbf{X} \tag{3.7.10}$$

Again, by carrying the derivation through we find that the error will propagate with modes that are the same as for the calculation itself.

PROBLEM 3.10

Redo the error derivation for a linear, multistate system described by

$$\dot{\mathbf{X}} = A\mathbf{X}$$

3.8 THE BACKWARD EULER ALGORITHM: AN EXAMPLE OF AN IMPLICIT INTEGRATION METHOD

As we have found, the basic Euler method suffers from several defects, including the possibility of instability. In this section we will examine a method that can be used to eliminate the problem of instability. This method is not as general as the straightforward Euler method, and is not entirely fault-free either, but it does provide us with one simple extension of the basic Euler method that is worthwhile pursuing.

The "backward Euler method" is very similar to the technique we have been using. The only difference is that we evaluate the state derivative at the new point, $(k + 1) DT$, instead of the old point, $k DT$. In other words, we use the formula

$$\mathbf{X}(t+DT) = \mathbf{X}(t) + DT\,\dot{\mathbf{X}}(t+DT) \tag{3.8.1}$$

There would seem to be some fundamental problems with this algorithm. Reading this carefully, we see that we need to evaluate the value of the derivative at $((k + 1) DT)$, and this will probably require that we know the state at $(k + 1) DT$. In other words, we need to know the value of the state before we can compute the state. This makes our computations more than just a little difficult, unless we can get out of this "philosophical" bind.

One possible way to get out of the bind is to use the "forward Euler" method (that is, our tried-and-true method of a few sections back) to compute a value for $x((k + 1) DT)$, and then to use that value to compute the value of the new derivative. That starts getting us into the area of algorithms generally called "predictor-corrector" algorithms, and we will have more to say about that type of algorithm later. However, in certain special systems, with special sorts of state equations, it may be possible to overcome our difficulty another way.

Let's look at a simple first-order linear differential equation, the type generated by applying KCL to the circuit we considered earlier:

$$\dot{x}(t) = ax(t) + bu(t) \tag{3.8.2}$$

where

$$x(t) = \text{system state and } u(t) = \text{system input.}$$

Now, applying the backward Euler formula to this system, we find

$$
\begin{aligned}
x((k+1)\,DT) &= x(k\,DT) + DT\,\dot{x}((k+1)\,DT) \\
&= x(k\,DT) + DT\,(ax((k+1)\,DT) + bu((k+1)\,DT)) \tag{3.8.3}
\end{aligned}
$$

This last equation can be solved for $x((k + 1) DT)$, giving

$$x((k+1)\,DT) = \frac{x(k\,DT) + bu((k+1)\,DT)}{(1 - a\,DT)} \tag{3.8.4}$$

What we have been able to do here is to take advantage of the specific form of the system to solve the backward Euler formula for the new value of the state in terms of the old state value and the new input value (which is assumed to be known). What we have to do yet is examine the stability and accuracy properties of this algorithm.

Assume that a is negative. Then we have

$$\frac{1}{1 - a\,DT} < 1 \qquad \text{for any positive } DT \tag{3.8.5}$$

This puts the pole inside the unit circle for this computation, and the computation is stable for any positive DT. This algorithm never exhibits the numerical instability associated with the forward Euler and the second-order Runge-Kutta.

Now let us see what happens in a linear system with more than one state. If we have

$$\dot{\mathbf{X}} = A * \mathbf{X} + \mathbf{B} * U \tag{3.8.6}$$

Then, when we do the computation, we get

$$\mathbf{X}(t + DT) = \mathbf{X}(t) + DT(A\mathbf{X}(t + DT) + \mathbf{B}U(t + DT)) \tag{3.8.7}$$

So we have

$$\mathbf{X}(t + DT) = (I - A * DT)^{-1} * (\mathbf{X}(t) + \mathbf{B} * U(t + DT)) \tag{3.8.8}$$

If we carry through a diagonalization of A, as we did earlier for the forward Euler, we will find that this algorithm is stable for any positive DT in the multiple state case also.

Of course, we still need to determine whether the method is accurate. Table 3.12 shows results for one step in a single-state system starting from a unit initial condition, with zero input:

$$\dot{x} = -x \qquad x(0) = 1 \tag{3.8.9}$$

As we can see, the backward Euler is capable of remaining stable at large multiples of a time constant, but it can become inaccurate. Still, if we have a large time constant ratio, it may be a reasonable algorithm for calculation of transients at multiples of the small time constant. The backward Euler also arises naturally in certain circuit contexts, and this is the topic of the next section.

3.9 COMPANION MODELS: ANOTHER CONCEPT FOR IMPLEMENTATION OF INTEGRATION ALGORITHMS

We have considered using an integration algorithm implemented with separate subroutines describing the algorithm and the system, and for control. There are other approaches to implementation of integration algorithms. In this section we will look at a different concept for using the algorithm.

Consider a capacitor, as illustrated in Figure 3.9. Imagine the capacitor in

TABLE 3.12

DT	Computed x(DT)	Actual x(DT)=EXP(−DT)
.100000E+00	.909091E+00	.904837E+00
.200000E+00	.833333E+00	.818731E+00
.500000E+00	.666667E+00	.606531E+00
.100000E+01	.500000E+00	.367879E+00
.200000E+01	.333333E+00	.135335E+00
.500000E+01	.166667E+00	.673795E−02
.100000E+02	.909091E−01	.453999E−04

Figure 3.9 An isolated capacitor.

isolation for now. Later, we will worry about the relation of this capacitor to a network in which it is embedded. For now, however, we can write the defining equation for a capacitor, relating voltage and current in the capacitor.

$$\dot{V}_c = \frac{I_c}{C} \qquad (3.9.1)$$

If we use the backward Euler algorithm to compute voltage for this capacitor, we will have

$$V_c((k+1)\,DT) = V_c(k\,DT) + DT\,\dot{V}_c((k+1)\,DT)$$
$$= V_c(k\,DT) + \left(\frac{DT}{C}\right) I_c((k+1)\,DT) \qquad (3.9.2)$$

This last expression can be solved for the current, $I_c((k+1)\,DT)$:

$$I_c((k+1)\,DT) = \left(\frac{C}{DT}\right) V_c((k+1)\,DT) - \left(\frac{C}{DT}\right) V_c(k\,DT) \qquad (3.9.3)$$

This expression for $I_c((k+1)\,DT)$ consists of two terms. The first term, $(C/DT)V_c((k+1)\,DT)$, is proportional to the voltage across the capacitor at $(k+1)\,DT$, the same time the current is computed for. The second term, $(C/DT)V_c(k\,DT)$, depends upon previously computed, known quantities. Both of these terms can be given a circuit interpretation. Refer to Figure 3.10. There each of these terms is identified with a current in a small equivalent circuit that satisfies the same relationships as we have above in mathematical form. Thus, we can interpret the backward Euler in these circuit terms, thinking of the $V_c(k\,DT)$ term as a known applied current source, and (C/DT) as a conductance in this equivalent circuit.

If we have a network composed of resistors, sources, and capacitors, we can compute the transient response by using this equivalent circuit.

To use the equivalent circuit we would replace all capacitors by their backward Euler equivalents for purposes of doing a transient analysis. The analysis would proceed by using all capacitor initial conditions to determine the starting

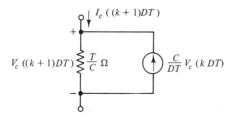

Figure 3.10 A circuit model of a capacitor using the backward Euler algorithm.

values of the current sources in all of the equivalent circuits. From this data (which is just all values for $V_c(0)$ for all capacitors) we can compute all the capacitor voltages at the next time step, and continue iterating after that.

PROBLEM 3.11

Compute the transient step response of the single time constant circuit of Figure 3.8 using the backward Euler and the method outlined just above.

The backward Euler can be used with other components. For the inductor we can generate a similar model. For an inductor, in general, we will have

$$\dot{I}_l = \frac{V_l}{L}$$

Then applying the backward Euler formula, we have

$$I_l((k+1)\,DT) = I_l(k\,DT) + \left(\frac{DT}{L}\right)V_l((k+1)\,DT) \qquad (3.9.4)$$

Again, we can give a circuit interpretation to the backward Euler relation. In Figure 3.12 we see an equivalent circuit that satisfies the same relations as the backward Euler formula for the inductor.

These equivalent circuits for capacitors and inductors can now be used within a linear circuit analysis program. If we do an interative analysis using these linear models, then we can do a complete simulation based on the backward Euler algorithm. The steps necessary to implement this approach are as follows:

1. Store the topological information describing the original circuit.
2. Replace capacitors and inductors by their equivalents for the backward Euler algorithm, using the initial conditions to determine the sources in the equivalents.
3. Do a linear analysis of the circuit.
4. Using the new conditions, update the sources and redo the linear analysis. Update the time variable. Repeat as often as needed to run through the desired time interval.

A program fragment to do this analysis is given in Appendix E. This subroutine can be transferred to whenever the user wants to do a transient analysis using constant sources.

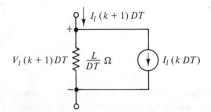

Figure 3.11 A circuit model of an inductor using the backward Euler algorithm.

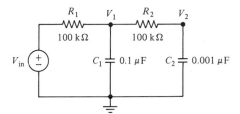

Figure 3.12 An example circuit.

This approach to doing transient analysis is relatively simple. Not much new coding is needed, and the programs we have developed in preceding chapters are easily modified to permit transient analysis capability.

In the next section we will examine a specific example of a transient analysis using backward Euler companion models.

3.10 AN EXTENDED EXAMPLE USING COMPANION MODELS

Consider the low-pass filter composed of two resistances and two capacitances as shown in Figure 3.12 (which is the same circuit given in Example 3.7). Replacing the capacitors by their equivalent companion models produces the circuit in Figure 3.13. In this network, we know that the time constants are widely different, with one time constant near 0.01 s and the other near 0.01 ms.

We are going to determine parameter values in the equivalent companion circuit using a time interval, DT, of 0.1 ms (10^{-4} s). Note that this is ten times the smallest time constant of 0.01 ms.

For the 0.1-μF capacitor, the companion model resistance and current source values are

$$R_{\text{comp1}} = \frac{DT}{C_1} = \frac{10^{-4}}{10^{-7}} = 1000 \ \Omega \qquad (3.10.1)$$

$$I_{\text{comp1}}(k \, DT) = 0.001 \, VC_1(k \, DT) \qquad (3.10.2)$$

Similarly, for C_2 we have

$$R_{\text{comp2}} = \frac{DT}{C_2} = \frac{10^{-4}}{10^{-9}} = 100{,}000 \ \Omega \qquad (3.10.3)$$

$$I_{\text{comp2}}(k \, DT) = 10^{-5} \, VC_2(k \, DT) \qquad (3.10.4)$$

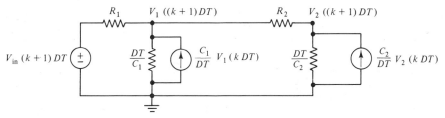

Figure 3.13 Companion models inserted into the example circuit of Figure 3.12.

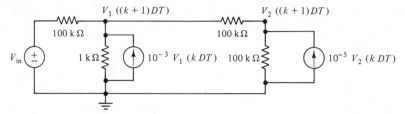

Figure 3.14 Numerical values inserted for the companion model of Figure 3.13.

This leads to the equivalent companion circuit shown in Figure 3.14. The node equations for this circuit are

$$\frac{V_1((k+1)\,DT) - V_{\text{in}}((k+1)\,DT)}{100{,}000} + \frac{V_1((k+1)\,DT) - V_2((k+1)\,DT)}{100{,}000}$$

$$+ \frac{V_1((k+1)\,DT)}{1000} = 0.001\,V_1(k\,DT) \qquad \text{(Node 1)}$$

$$\frac{V_2((k+1)\,DT) - V_1((k+1)\,DT)}{100{,}000} + \frac{V_2((k+1)\,DT)}{100{,}000} = 10^{-5}V_2(k\,DT) \qquad (3.10.5)$$

These scalar equations can be rearranged to give

$$\begin{bmatrix} 1.02 & -0.01 \\ -0.01 & 0.02 \end{bmatrix} * \begin{bmatrix} V_1((k+1)\,DT \\ V_2((k+1)\,DT \end{bmatrix} = \begin{bmatrix} 0.01\,V_{\text{in}}((k+1)\,DT) + V_1(k\,DT) \\ (0.02)V_2(k\,DT) \end{bmatrix}$$

$$(3.10.6)$$

Taking the inverse of the matrix, we find

$$\begin{bmatrix} V_1((k+1)\,DT \\ V_2((k+1)\,DT \end{bmatrix} = \frac{1}{0.0203} * \begin{bmatrix} 0.02 & 0.01 \\ 0.01 & 1.02 \end{bmatrix} * \begin{bmatrix} 0.01\,V_{\text{in}}((k+1)\,DT) + V_1(k\,DT) \\ (0.02)V_2(k\,DT) \end{bmatrix}$$

$$(3.10.7)$$

PROBLEM 3.11

Write a short program to do the computations to compute V_1 and V_2 when V_{in} is a constant 10 V.

3.11 GENERALIZATIONS AND CONCLUSIONS

The integration algorithms presented in this chapter have the property that they match terms in a Taylor series:

$$x(t + DT) = x(t) + \dot{x}(t)\,DT + \ddot{x}(t)\frac{(DT)^2}{2} + \dddot{x}(t)\frac{(DT)^3}{6} + \cdots \qquad (3.11.1)$$

The forward Euler uses the first two terms and neglects the rest. The second-order Runge-Kutta produces a final calculation that matches through the $(DT)^2$

term, effectively ignoring the rest. (The backward Euler algorithm is not quite as simple as these.) If the simulation results are close to the true values, then it is reasonable to assume that the value of DT chosen is small enough that the first neglected term is small, but that it is larger than the other neglected terms. The first neglected term is a reasonable estimate of the error made in a single integration step. These terms are as follows:

$$\ddot{x}(t)\frac{(DT)^2}{2} \qquad \text{Euler} \qquad\qquad (3.11.2)$$

$$\dddot{x}(t)\frac{(DT)^3}{6} \qquad \text{Second-order R-K} \qquad\qquad (3.11.3)$$

Clearly, this "single-step error" can be controlled by choosing a small enough value for T. For the Euler algorithm the error will be approximately proportional to $(DT)^2$. For the second-order Runge-Kutta, the error is approximately proportional to $(DT)^3$. The fact that the error depends upon a power or the integration interval, DT, means that reducing the integration interval can pay significant dividends in accuracy. Moreover, the improvement in accuracy should be better for the higher-order algorithm because of the higher power involved.

One obvious question raised by these considerations is whether or not it would pay to match further terms in the Taylor series. The popular fourth-order Runge-Kutta integration algorithm does just that, matching terms up to $(DT)^4$, with an attendant error proportional to $(DT)^5$. The price paid with this algorithm is that the derivative function must be evaluated four times for each integration step. However, despite these seeming disadvantages, the fourth-order Runge-Kutta is an extremely widely used integration method. We will delve more deeply into this method and its properties in the next chapter.

Another unresolved issue in this area is that there is an accumulation of error in any integration calculation. The errors just considered are the errors that crop up in a single step. Once an error is made, it is carried along in subsequent integration steps. That error could grow or die out. In this chapter we haven't addressed that issue, but it seems clear that we have to worry about accumulated error.

One common way of handling error is very "ad hoc." Many times a simulation is run, then all or part of the simulation is done over using a smaller integration interval. If the results are close enough to satisfy the user, then it is assumed to be accurate. Obviously, this depends very much on the particular situation.

Finally, in this chapter we introduced the idea of using a linear circuit model for capacitors and inductors whenever the backward Euler algorithm is used. The backward Euler is an example of an implicit algorithm, and again we have opened up a topic that deserves further treatment. Again, in the next chapter, we will look into implicit algorithms and the backward Euler, particularly, in some more detail. These algorithms have the potential to do calculations at large time-constant multiples, and are potentially valuable.

REFERENCES

1. P. M. DeRusso, R. J. Roy and C. M. Close, *State Variables for Engineers,* Wiley, New York, 1965. (This book is a very well-written introduction to state variables, and can be considered a classic in the field.)
2. S. W. Director and R. A. Rohrer, *Introduction to Systems Theory,* McGraw-Hill, New York, 1972. (see especially pages 124–130.)
3. T. M. Apostol, *Mathematical Analysis,* Addison-Wesley, Reading, MA, 1957, p. 417.
4. J. E. Gibson, *Nonlinear Automatic Control,* McGraw-Hill, New York, 1963.
5. Robert M. May, *Stability and Complexity in Model Ecosystems,* Princeton University Press, Princeton, NJ, 1973.
6. R. P. Prince, W. Giger, Jr., J. W. Bartok, Jr., and T. L. Logee, "Controlled Environment Plant Growth," A Report Submitted to the Environment Committee of the General Assembly, State of Connecticut. Department of Agricultural Engineering, University of Connecticut, Storrs, CT 06268, April 1976.

BIBLIOGRAPHY

L. O. Chua and P-M Lin, *Computer Aided Analysis of Electronic Circuits: Algorithms and Computational Techniques,* Prentice-Hall, Englewood Cliffs, NJ, 1975, pp. 46–50.

Bernard C. Patten (Ed.), *Systems Analysis and Simulation in Ecology,* vol. 1, Academic Press, New York, 1971.

MINIPROJECTS

1. The block diagram of a system with two time constants ($1/a$ and $1/b$) is shown. The state equations for this system are

$$\dot{x}_1 = -ax_1 + ax_2$$
$$\dot{x}_2 = -bx_2 + bu$$

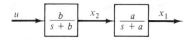

Consider three numerical cases:

$$b = 1.0 \qquad a = 10$$
$$b = 1.0 \qquad a = 500$$
$$b = 10 \qquad a = 1.0$$

Using both the Euler algorithm and the second-order Runge-Kutta, determine stability limits for the integration interval, and determine when the accuracy of the calculation stays within 0.1 V for a 10-V step intput.

2. An interesting set of nonlinear differential equations comes out of a simple model for the population dynamics of two interacting biological species. A particular example

is discussed by Gibson [4] in a problem describing the interaction of sharks and sole. The sharks prey on the sole, and a set of differential equations that describe the dynamic situation is as follows:

$$\dot{x}_1 = 2x_1 - x_1 x_2$$
$$\dot{x}_2 = -8x_2 + 2x_1 x_2$$

where x_1 = sole population and x_2 = shark population. It is known that the solution of these equations oscillates around the steady-state solution, and that these oscillations do not decay. It is also known that a good test of how well an integration algorithm behaves is how well it can integrate oscillatory systems without decaying or growing. Investigate the behavior of the Euler and second-order Runge-Kutta for this system and a linear oscillatory system of your choice.

Further reading (and more numerical examples) is available in a text by May [5], especially chap. 2.

3. Using a companion model:
   (a) Derive the set of equations that must be solved repetitively to produce a numerical solution for the circuit in Figure 3.7.
   (b) Program the result of part (a).
   (c) Evaluate the stability properties of this method and compare with the forward Euler method.

4. (a) Derive the difference equation that results from applying the backward Euler to a linear constant-coefficient system with inputs

$$\dot{X} = AX + Bu$$

   (b) Program your result.
   (c) Use the program on the three example systems given in Problem 1.

5. The circuit shown has been simulated using our interactive program; here, $L = 1$ H and

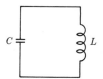

$C = 1000\ \mu F$. For the simulation, we used $DT = 0.005$ s, but printed out at 0.01-s intervals. Although the system has no resistance, the simulation results show the oscillations drying out. Check the behavior of this system when the backward Euler method is used in the simulation. Calculate the *energy* stored in the system at each integration step and determine whether the energy increases or decreases. Are the simulation results reasonable for the values used? Compare with what happens using a *forward* Euler algorithm (by computing energy at every step).

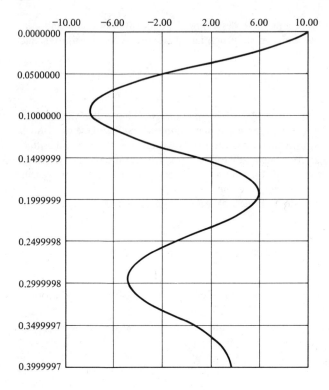

Chapter 4

Simulation II

4.1 INTRODUCTION

Simulation calculations can take large amounts of CPU time if the simulated system is "stiff." For example, if the system is linear and the eigenvalues of the A matrix have large discrepancies in size, the system would be stiff. Similar effects occur in nonlinear systems. Some authors have even defined a stiffness ratio as the ratio of the longest and shortest time constants (see Ralston and Rabinowitz [1]).

A great deal of effort has been expended in attempts to overcome the problem of excessive numbers of iterations in simulation of stiff systems. At least two approaches are worth considering. The simulation algorithms developed in the last chapter use calculations that match exact solutions to some given power of $DT,$ the integration interval. One approach is to attempt to include higher-order terms in the series solution to get more accuracy at each step or to permit taking larger steps. This approach yields higher-order Runge-Kutta methods (which are possibly the most widely used integration algorithms) and other algorithms like the predictor-corrector algorithm.

The second class of algorithms is the class of implicit integration algorithms. In Chapter 3 we briefly encountered the backward Euler algorithm, which is a representative of the implicit integration algorithms. These algorithms are probably the most effective for dealing with stiff systems of differential equations, but they do present interesting and awkward numerical problems in implementation.

In this chapter we will discuss both of these approaches, starting with higher-order methods, then later considering implicit methods. We will discuss some of the numerical problems of implementation of implicit algorithms. Finally, we will examine a problem that requires more sophistication to get a good numerical solution.

4.2 HIGHER-ORDER RUNGE-KUTTA ALGORITHMS

The Euler algorithm computes a solution that matches the true solution whenever the true solution has a constant derivative. Whenever the state derivatives are not constant, we will have an error as shown in Figure 4.1. There are numerous ways to improve this situation, other than just reducing the integration interval. Perhaps the most obvious way to improve accuracy is to include higher-order terms in the expression for the computed value of $X(t + DT)$:

$$X(t+DT) = X(t) + \dot{X}(t)\,DT + \ddot{X}(t)\frac{(DT)^2}{2} \qquad (4.2.1)$$

where

$$\dot{X}(t) = \text{First derivative of } X(t), \ldots \qquad (4.2.2)$$

We would like to consider how we might be able to devise an algorithm that effectively includes more terms in the Taylor series. The problem we have to overcome is that we can rather easily compute *first* derivatives using the state equations, but it gets difficult computing higher-order derivatives with respect to time if we have only the state equations. One way around this difficulty is to use multiple evaluations of the first derivative at different points, instead of calculating higher-order derivatives. It isn't obvious that using multiple first-derivative evaluations will permit us to evade calculating higher-order derivatives, but in fact the technique can be made to work.

The second-order Runge-Kutta algorithm is the simplest instance of this approach. We can give a development for $x(t + DT)$ that includes quadratic terms in the time increment, DT. As we can see, the final expression involves $(DT)^2$, and second (partial) derivatives of the derivative function, $f(x, t)$. That is,

$$
\begin{aligned}
x(t+DT) &= x(t) + DT\,\frac{dx(t)}{dt} + \frac{(DT)^2}{2}\frac{d^2x(t)}{dt^2} + \cdots \\
&= x(t) + DT\,f(x,t) + \frac{(DT)^2}{2}\left[\frac{\partial f(x,t)}{\partial t} + \frac{\partial f(x,t)}{\partial x}\frac{dx(t)}{dt}\right] + \cdots \\
&= x(t) + DT\,f(x,t) + \frac{(DT)^2}{2}\left[\frac{\partial f(x,t)}{\partial t} + \frac{\partial f(x,t)}{\partial x}f(x,t)\right] + \cdots \qquad (4.2.3)
\end{aligned}
$$

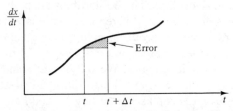

Figure 4.1   The error in the Euler integration algorithm.

The starting point for our argument is the single nonlinear state equation for a single state:

$$\frac{dx}{dt} = f(x, t) \qquad (4.2.4)$$

Now, if we are going to compute an approximate value for $x(t + DT)$, we are almost forced to start by computing the derivative $f(x, t)$. Once we have done that derivative computation, we can estimate the new value, $x(t + DT)$:

$$x(t + DT) \simeq x(t) + DT f(x, t) \qquad (4.2.5)$$

Of course, this is just what we get using the Euler algorithm. However, now we use this new value to compute a new derivative value. Let $k_1 = f(x, t)$ be the result of our first-derivative computation. Then, we evaluate a new derivative, calling it k_2:

$$k_2 = f((x(t) + DT\, k_1), t + DT) \qquad (4.2.6)$$

Then, with these two computed derivative values, we can get a more refined estimate for $x(t + DT)$. As an approximation for $x(t + DT)$ let us take the expression

$$x_{\text{calc}}(t + DT) = x(t) + DT(ak_1 + bk_2) \qquad (4.2.7)$$

This is just a weighted sum of the two computed derivative values, k_1 and k_2. It seems reasonable that these weights would have to sum to unity if this scheme is going to work, so we should have

$$a + b = 1 \qquad (4.2.8)$$

If we now write out the detailed expressions for k_1 and k_2 and substitute in our final expression, we have the development presented next:

$$
\begin{aligned}
x_{\text{calc}}(t + DT) &= x(t) + [ak_1 + bk_2]\, DT \\
&= x(t) + a\, DT\, f(x, t) \\
&\quad + b\, DT\left[f(x, t) + \frac{\partial f}{\partial x} - DT f(x(t), t) + DT\frac{\partial f(x, t)}{\partial t} + \cdots \right] \\
&= x(t) + DT[a + b] f(x, t) + b(DT)^2 \\
&\quad \times \left[\frac{\partial f(x, t)}{\partial x} f(x, t) + \frac{\partial f(x, t)}{\partial t} \right] + \cdots
\end{aligned}
\qquad (4.2.9)
$$

In the final expression, we can make both the linear and quadratic terms match those in the Taylor series by requiring

$$a + b = 1 \quad \text{and} \quad b = \frac{1}{2} \qquad (4.2.10)$$

and hence, also

$$a = \frac{1}{2} \qquad (4.2.11)$$

This produces a second-order Runge-Kutta integration algorithm. Runge and Kutta [2, 3] had the original ideas. It is a second-order algorithm because it matches terms out to the second derivative in the Taylor series expansion. Higher-order Runge-Kutta are possible; the fourth-order Runge-Kutta algorithm is probably the most widely used integration algorithm.

ALGORITHM

SECOND-ORDER RUNGE-KUTTA

The second-order Runge-Kutta integration algorithm is used to solve sets of first-order differential equations. The algorithm computes solutions that are locally accurate to quadratic terms in the Taylor series expansion about the last computed point. Steps in the algorithm are as follows:

1. Assume some given value for the states, $\mathbf{X}(t)$.
2. Compute an estimate of the change in $\mathbf{X}(t)$ using the given starting value for $\mathbf{X}(t)$:

$$\mathbf{X}(t+DT)_{\text{est}} = \mathbf{W}_1 = \mathbf{X}(t) + DT\,\mathbf{f}(\mathbf{X}(t),t) = \mathbf{X}(t) + \mathbf{K}_1$$

3. Compute a second estimate change, using the first estimate computed in part B to form the derivative:

$$\mathbf{K}_2 = DT\,\mathbf{f}(\mathbf{W}_1, t+DT)$$

4. Form the computed value for $\mathbf{X}(t+DT)$ as a weighted average of the two estimated changes of steps 2 and 3:

$$\mathbf{X}_{\text{calc}}(t+DT) = \mathbf{X}(t) + \frac{\mathbf{K}_1 + \mathbf{K}_2}{2}$$

Table 4.1 shows one implementation of a second-order Runge-Kutta in FORTRAN. To use this algorithm in our previous programs would require just deleting the call on EULER, and substituting a call on RUNGE2 instead.

We should note that this derivation does not give the most general second-order Runge-Kutta, and that the algorithm could take other forms. For example, it is not necessary to do the second-derivative evaluation at $t + DT$; it could be done at $t + (DT/2)$, and the results still used to estimate $x(t + DT)$. More general forms of the algorithm are derived in several of the references [1, 4]. The particular second order Runge-Kutta described in the preceding material is discussed and derived in Scraton [5]. It travels under the pseudonym of "Heun's algorithm" or "modified trapezoidal algorithm" [6].

Another family of second-order Runge-Kutta algorithms can be obtained by doing evaluations of derivatives within the interval from t to $t + DT$. One of these involves a midinterval evaluation of derivatives, leading to the expression

$$\mathbf{K}_1 = \frac{DT}{2}\mathbf{f}(\mathbf{X}(t), t) \qquad\qquad (4.2.12a)$$

TABLE 4.1

```
C
C*****************************************************
C
      SUBROUTINE RUNGE2(X,T,DT,N)
      DIMENSION X(10), K1(10), K2(10)
      REAL K1,K2
C
C THIS IS A SECOND ORDER RUNGE-KUTTA ALGORITHM
C
      CALL XDOT (X,K1,T,N)
      DO 20 I = 1,N
   20 XTEMP(I) = X(I) + T*K1(I)
      T = T + DT
      CALL XDOT (XTEMP,K2,T,N)
      DO 100 I = 1,N
  100 X(I) = X(I) + (K1(I) + K2(I))/2.
      RETURN
      END
C
C*****************************************************
C
```

$$\mathbf{K}_2 = DT\, \mathbf{f}\!\left(\mathbf{X}\!\left(t + \frac{DT}{2}\right), t + \frac{DT}{2}\right) \qquad (4.2.12b)$$

$$\mathbf{X}(t+DT)_{\text{calc}} = \mathbf{X}(t) + \mathbf{K}_2 \qquad (4.2.12c)$$

This algorithm is referred to as the "modified Euler-Cauchy algorithm" [6] or as the "midpoint method" [7].

PROBLEM 4.1

Simulate a first-order single-time-constant system using the second-order Runge-Kutta. Experimentally determine the largest ratio of integration interval to time constant that produces stable calculations.

Runge-Kutta algorithms essentially match terms with terms in the Taylor series of the actual solution. Different-order Runge-Kutta algorithms match different-order terms. The fourth-order Runge-Kutta (which matches terms up to the fourth power of DT) has proved to be a simulation algorithm (for example, the fourth-order Runge-Kutta is the default integration algorithm in the simulation language ACSL [8]).

As in all Runge-Kutta algorithms, the fourth-order algorithm uses several (four) first-derivative evaluations as a way of evading higher-order derivative evaluation. The expressions involved are

$$\mathbf{K}_1 = \mathbf{f}(\mathbf{X}(t), t)$$

$$\mathbf{K}_2 = \mathbf{f}\!\left(\mathbf{X}(t) + \frac{DT}{2}\mathbf{K}_1, t + \frac{DT}{2}\right)$$

$$\mathbf{K}_3 = \mathbf{f}\left(\mathbf{X}(t) + \frac{DT}{2}\mathbf{K}_2, t + \frac{DT}{2}\right)$$

$$\mathbf{K}_4 = \mathbf{f}(\mathbf{X}(t) + DT\,\mathbf{K}_3, t + DT)$$

$$\mathbf{X}_{calc}(t + DT) = \mathbf{X}(t) + \frac{DT}{6}[\mathbf{K}_1 + 2\mathbf{K}_2 + 2\mathbf{K}_3 + \mathbf{K}_4] \qquad (4.2.13)$$

A FORTRAN implementation (consistent with other routines presented earlier) is given in Table 4.2.

ALGORITHM

FOURTH-ORDER RUNGE-KUTTA

The fourth-order Runge-Kutta integration algorithm is used to integrate sets of first-order differential equations. The method computes solutions accurate to terms in the fourth power of DT by doing repetitive evaluations of the first derivatives. It does not need expressions for second, third, or fourth derivatives.

The steps in the algorithm are as follows:

1. Assume a starting state vector, $\mathbf{X}(t)$.
2. Successively calculate $\mathbf{K}_1$, $\mathbf{K}_2$, $\mathbf{K}_3$ and $\mathbf{K}_4$:

$$\mathbf{K}_1 = \mathbf{f}(\mathbf{X}(t), t)$$

$$\mathbf{K}_2 = \mathbf{f}\left(\mathbf{X}(t) + \frac{DT}{2}\mathbf{K}_1, t + \frac{DT}{2}\right)$$

TABLE 4.2

```
      SUBROUTINE RUNGE4 (X,T,DT,N)
      DIMENSION X(10), XD(10), Y1(10), Y2(10), Y3(10)
C
C RUNGE4 IS A FOURTH ORDER RUNGE KUTTA INTEGRATION ROUTINE
C RUNGE4 PERFORMS A SINGLE STEP INTEGRATION
C SEE JAMES, SMITH AND WOLFORD   PP345-356
C
C N FIRST ORDER DIFFERENTIAL EQUATIONS ARE INTEGRATED.
C DT IS THE TIME INTERVAL BETWEEN STEPS
C T IS THE TIME
C X IS AN ARRAY CONTAINING THE STATES THAT ARE TO BE INTEGRATED
C NEW VALUES OF X ARE CALCULATED AND RETURNED IN X
C
      DT2=DT/2.
      CALL XDOT (X,XD,T,N)
      DO 210 I=1,N
  210 Y1(I)=DT2*XD(I)+X(I)
      T=T+DT/2.
      CALL XDOT (Y1,XD,T,N)
      DO 220 I=1,N
  220 Y2(I)=DT2*XD(I)+X(I)
      CALL XDOT (Y2,XD,T,N)
      DO 230 I=1,N
  230 Y3(I)=DT*XD(I)+X(I)
      T=T+DT/2.
      CALL XDOT (Y3,XD,T,N)
      DO 240 I=1,N
  240 X(I)=(2.*Y1(I)+4.*Y2(I)+2.*Y3(I)+DT*XD(I)-2.*X(I))/6.
      RETURN
      END
```

$$\mathbf{K}_3 = \mathbf{f}\left(\mathbf{X}(t) + \frac{DT}{2}\mathbf{K}_2, t + \frac{DT}{2}\right)$$

$$\mathbf{K}_4 = \mathbf{f}(\mathbf{X}(t) + DT\mathbf{K}_3, t + DT)$$

3. Calculate $\mathbf{X}(t, DT)$

$$\mathbf{X}_{\text{calc}}(t + DT) = \mathbf{X}(t) + [\mathbf{K}_1 + 2\mathbf{K}_2 + 2\mathbf{K}_3 + \mathbf{K}_4]\frac{DT}{6}$$

4. Cycle through steps 2 and 3, incrementing t by DT until maximum time is reached.

Before we leave the topic of Runge-Kutta methods, we should note their error behavior. Generally, an Nth-order Runge-Kutta will have an error that depends upon the $(N + 1)$th power of the integration interval. Thus, the fourth-order Runge-Kutta will have an error that depends upon the fifth power of DT. One implication of that behavior is that once we determine a stable range for DT, small decreases in DT can produce much more accurate computations.

The important aspect of Runge-Kutta behavior, however, is best appreciated by working through Miniproject 2. Increasing the order of the algorithm does not produce order-of-magnitude changes in the stable region, and most Runge-Kutta methods will have difficulty with stiff sets of differential equations.

PROBLEM 4.2

A pump is used to transfer a liquid into a tall, narrow cylinder at a constant rate of 2 liters per minute. The cylinder has an orifice at the bottom that allows the liquid to drain out. The rate of flow through the orifice is proportional to the square root of the pressure. The pressure in turn depends upon the height of the liquid column in the cylinder, so we have

$$P = \text{Pressure (g/cm}^2)$$
$$\rho = \text{Fluid density (g/cm}^3)$$
$$h = \text{Height (cm)}$$
$$A = \text{Cross-sectional area} = 4 \text{ cm}^2$$
$$\text{Inflow rate} = 2 \text{ liters/min}$$
$$\text{Outflow rate} = KP^{1/2} = KP^{1/2}h^{1/2}$$
$$\text{Rate of accumulation} = \text{Inflow} - \text{outflow, or}$$
$$A\frac{dh}{dt} = 2 - \sqrt{2}h^{1/2}$$

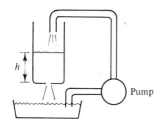
Pump

This final first-order differential equation can be integrated using any of the integration algorithms presented in Chapters 3 and 4.

(a) Using the Euler integration algorithm, determine the time it takes for the liquid level to reach 0.25 m and to reach 0.5 m.

(b) Repeat using the second-order Runge-Kutta algorithm.

In using each algorithm, you will have to determine a range of integration interval that gives accurate answers. Assume that you want to know the time interval to within 0.05 s. Determine the integration interval for each algorithm and compare the number of function evaluations needed in each case. Is there any advantage to using the second-order Runge-Kutta? (For some background on this problem, see McClamroch [9].)

PROBLEM 4.3

Models of population growth and decay are interesting nonlinear systems. A single species will grow until it reaches the largest size that its food supply will support. A simple model for the growth of a population P is given by the differential equation

$$\frac{dP}{dt} = rP\left(1 - \frac{P}{K}\right)$$

In this model, when P is much smaller than K, the differential equation is approximately $dP/dt = rP$, which has a solution with exponential growth. As the population gets larger, the second factor, $(1 - P/K)$, acts to decrease the growth rate. If, at any time, the population reaches the saturation level K, then the growth rate goes to zero and growth ceases. Thus, the population can never "overshoot" the limiting level K.

A numerical solution to this differential equation could take on a value larger than K. Solve the differential equation using the second-order Runge-Kutta.

(a) Assume a starting population of 500, a limiting population of 10,000, and a growth constant r of 0.2. Determine the highest value of DT for which no overshoot occurs in the numerical solution.

(b) This differential equation can be linearized near the limiting population level, K:

$$P = K + \Delta P$$

$$\frac{dP}{dt} = \frac{d\,\Delta P}{dt} = r(K + \Delta P)\left[1 - \left(\frac{K + \Delta P}{K}\right)\right]$$

$$= -r\,\Delta P - \frac{r(\Delta P)^2}{K}$$

$$\frac{d\,\Delta P}{dt} \simeq -r\,\Delta P$$

From this linearization we would expect the numerical solution to behave like a single time constant problem, as in Problem 4.1. Compare your results with the results of Problem 4.1 to determine whether the behavior in the population problem near the limit has the same numerical behavior as the linear single-time-constant problem.

4.3 PREDICTOR-CORRECTOR ALGORITHMS

The algorithms discussed so far are "one-point" algorithms. Given an initial state it is possible to apply the algorithm and compute a state for a time one integration interval later. In many instances a "multipoint" algorithm would be preferable. For example, approximating the integration area by a trapezoid is one way of getting more accuracy. Figure 4.2 illustrates the difference in error when we can evaluate the derivative at two points instead of one.

The problem with this trapezoidal method of integration is fairly obvious. We need to know $x(t + DT)$ in order to evaluate the derivative, $f(x(t + DT), t + DT)$, but we need to know the derivative in order to compute $x(t + DT)$. This "circular" type of computation comes about because the desired quantity, $x(t + DT)$, appears only implicitly in these expressions, and it may not always be possible to solve for $x(t + DT)$ explicitly.

One appealing strategy is to "estimate" $x(t + DT)$ and use this rough estimate to calculate the derivative at $(t + DT)$. This explicitly gets around calculations for $x(t + DT)$. The sequence of operations we are suggesting is as follows:

1. Use some known algorithm to calculate an estimate of $x(t + DT)$. For example, we could say

$$x_{est}(t + DT) \simeq x(t) + DT\, f(x(t), t) \qquad (4.3.1)$$

2. Then calculate the final version of $x(t + DT)$ as

$$x(t + DT) = x(t) + \frac{DT}{2}(f(x(t), t) + f(x_{est}(t + DT), t + DT)) \qquad (4.3.2)$$

This algorithm is probably the simplest possible predictor-corrector method. The name "predictor-corrector" comes about because we first "predict"

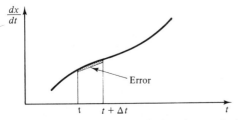

Figure 4.2 The trapezoidal integration method (and error).

$x(t + DT)$ (that is, we calculate $x_{est}(t + DT)$) and then we "correct" our estimate by doing another computation.

As we have done with other algorithms, we can see what transpires when we use a predictor-corrector on a simple first-order linear system. Consider the system

$$\frac{dx}{dt} = ax(t) \tag{4.3.3}$$

Then we have

$$x_{est}(t + DT) = x(t) + DT\, ax(t)$$
$$= (1 + a\, DT)x(t) \tag{4.3.4}$$

Our final calculation for $x(t + DT)$ is then

$$x_{calc}(t + DT) = x(t) + \frac{DT}{2}[x(t) + \{1 + a\, DT\}x(t)]$$

$$= \frac{1 + a\, DT + ((a\, DT)^2)}{2} x(t) \tag{4.3.5}$$

This computed solution is accurate to terms in $(DT)^2$, so the error is going to depend upon the cube of DT. Thus, the performance of this algorithm should approximate that of the second-order Runge-Kutta, at least as far as error is concerned.

There is another interesting property of the predictor-corrector concept. Once we think of the idea of correcting, it is not too hard to get to the idea of multiple corrections. In other words, we can use the corrected value of $x(t + DT)$ to produce a further corrected value, and then continue this process.

PROBLEM 4.4

Compute a second correction for the single-time-constant linear system just discussed. Is the solution correct to terms in $(DT)^3$?

This example of a predictor-corrector algorithm is just that, one example. There are many other algorithms in this family, and some of them are widely used. Ralston and Rabinowitz [1] probably have the best, most accessible discussion of predictor-corrector methods.

ALGORITHM

PREDICTOR-CORRECTOR INTEGRATION ALGORITHMS

Predictor-correctors are two-step algorithms. The first step does some sort of estimation of $X(t + DT)$, and in the second step that estimate is used to refine the computed value of $X(t + DT)$. The general steps in the algorithm are:

1. Compute an estimate, $X_{pred}(t + DT)$, of the value of $X(t + DT)$.
2. Using $X_{pred}(t + DT)$, compute a "corrected" version of $X(t + DT)$.

There are various differing versions of the predictor-corrector algorithm.
Euler algorithm:

$$\mathbf{X}_{pred}(t + DT) = \mathbf{X}(t) + \mathbf{F}(\mathbf{X}(t), t)\,DT \qquad \text{(Euler)}$$
$$\mathbf{X}_{corr}(t + DT) = \mathbf{X}(t) + \mathbf{F}(\mathbf{X}_{pred}(t + DT), t + DT)\,DT$$

Adams-Moulton algorithm:

$$\mathbf{X}_{pred}(t + DT) = \mathbf{X}(t) + \frac{DT}{24}[55\mathbf{F}(\mathbf{X}(t), t) - 59\mathbf{F}(\mathbf{X}(t - DT), t - DT)$$
$$+ 37\mathbf{F}(\mathbf{X}(t - 2\,DT), t - 2\,DT) - 9\mathbf{F}(\mathbf{X}(t - 3\,DT), t - 3\,DT)]$$
$$\mathbf{X}_{corr}(t + DT) = \mathbf{X}(t) + \frac{DT}{24}[9\mathbf{F}(\mathbf{X}_{pred}(t + DT), t + DT) + 19\mathbf{F}(\mathbf{X}(t), t)$$
$$- 5\mathbf{F}(\mathbf{X}(t - DT), t - DT) + \mathbf{F}(\mathbf{X}(t - 2\,DT), t - 2\,DT)]$$

We should note that many predictor-corrector algorithms are not self-start-
ing. In other words, the Adams-Moulton expressions given in the algorithm
description involve derivative evaluations for several times previous to the present
time. Initially, we will not have available all of the data we need to evaluate
those derivatives. If we start with initial values of the states, say at time $t = 0$,
then normally we would not need to know and would not specify values of the
states for $t = DT$, $t = -2\,DT$, and so on. That presents problems when we try
to start doing Adams-Moulton or other predictor-corrector calculations.

There are two solutions to this problem. First, we can use predictor-corrector
algorithms that do not have start-up problems. The Heun algorithm shown in
the algorithm development has no need for start-up values. It is also the least
accurate predictor-corrector. Modifying even the predictor expression to the following one can cause start-up problems and simultaneously improve accuracy
significantly:

$$\mathbf{X}_{calc}(t + DT) = \mathbf{X}(t) + \frac{DT}{2}[3\mathbf{F}(\mathbf{X}(t), t) - \mathbf{F}(\mathbf{X}(t - DT), t - DT)] \qquad (4.3.6)$$

The other way out of this dilemma is to use some other algorithm to
generate start-up values. Runge-Kuttas are often used for that purpose.

In any event, before proceeding, we need to note that the derivatives shown
in the expressions for the predictor-corrector algorithms do present those start-
up problems, and in the expressions for the Adams-Moulton (for example) the
derivatives indicated should use the best calculations possible for the states indicated.

Reflecting on what takes place in a predictor-corrector, we might realize
that we are really solving the nonlinear equation in $\mathbf{X}(t + DT)$ using functional
iteration. (We discussed functional iteration in Chapter 2 as a technique for
solving nonlinear sets of equations.) However, functional iteration is not necessarily the most efficient way of solving a set of nonlinear equations. Other
methods include, for example, the Newton-Raphson algorithm. Implicit methods
that lead us into using the Newton-Raphson algorithm are the subject of the
next section.

4.4 THE BACKWARD EULER: AN EXAMPLE OF AN IMPLICIT ALGORITHM

In Chapter 3 we discussed the backward Euler algorithm in a limited context. There we assumed that the system was linear and generated some specific methods tuned to linear time-invariant systems. Here, we want to extend the discussion to include more general nonlinear systems.

Assume that we are dealing with a system described by a set of differential equations:

$$\frac{d\mathbf{X}}{dt} = \mathbf{F}(\mathbf{X}, \mathbf{U}, t) \tag{4.4.1}$$

where

$$\mathbf{X}(t) = \text{State vector}$$
$$\mathbf{U}(t) = \text{Vector of inputs}$$
$$t = \text{Time}$$

Then the expression for the calculated value of $\mathbf{X}(t + DT)$, using the backward Euler algorithm for the calculation, is

$$\mathbf{X}(t+DT) = \mathbf{X}(t) + DT\,\mathbf{F}(\mathbf{X}(t+DT), \mathbf{U}(t+DT), t+DT)$$

Here, $\mathbf{X}(t + DT)$ appears on both sides of the equation, and is embedded in the function $\mathbf{F}(\ \ )$ on the right-hand side of the equation. Thus $\mathbf{X}(t + DT)$ is not given explicitly by this expression, but appears implicitly, and we must solve for $\mathbf{X}(t + DT)$.

In solving for $\mathbf{X}(t + DT)$, we are free to choose any solution method we like. If the system is linear, we can derive an explicit expression for $\mathbf{X}(t + DT)$. If the system is nonlinear, we are forced to use an iterative technique for the solution. The choice of solution method is not trivial. Chua and Lin [6] point out that a version of the Newton-Raphson algorithm is the most appropriate one here.

If we have a value for $\mathbf{X}(t)$, then the next computed value, $\mathbf{X}(t + DT)$ is the solution of

$$\mathbf{X}(t+DT) = \mathbf{X}(t) + DT * \mathbf{F}(\mathbf{X}(t+DT), \mathbf{U}(t+DT), t+DT) \tag{4.4.2}$$

The state derivative function, $\mathbf{F}(\mathbf{X}(t), \mathbf{U}(t), t)$ can be replaced by a linear approximation, linearizing around $\mathbf{X}(t)$, the last computed value. Doing the linearization we get

$$\begin{aligned}\mathbf{X}(t+DT) = \mathbf{X}(t) + DT[\mathbf{F}(\mathbf{X}(t), \mathbf{U}(t), t)\\ + J(\mathbf{X}(t), \mathbf{U}(t), t) * (\mathbf{X}(t+DT) - \mathbf{X}(t))]\end{aligned} \tag{4.4.3}$$

Solving for $\mathbf{X}(t + DT)$, we obtain

$$\mathbf{X}(t+DT) = \mathbf{X}(t) + [I - DT * J(\mathbf{X}(t), \mathbf{U}(t), t)]^{-1} DT * \mathbf{F}(\mathbf{X}(t), \mathbf{U}(t), t) \tag{4.4.4}$$

With this expression we have the basic iterative scheme that can be used to step along the backward Euler algorithm.

There are a few items that must be noted. First, each step in this algorithm requires a matrix inversion. That inversion will almost certainly take a large amount of computation time, and the cost of that computation time must be balanced against the possible increase in speed of computation made possible by using larger values of DT.

Second, we need to recognize that we have proposed doing only one iteration of the Newton-Raphson algorithm at each step. We just might be able to get away with that if $X(t)$ does not change that much with each iteration or if the process does not deviate far from being linear.

Finally, we need to recognize that the Jacobian must be evaluated numerically. Since the Jacobian is a matrix of derivatives, the process of calculating the Jacobian will be "noisy" of necessity.

To compute the Jacobian we must essentially calculate the derivatives of the expressions for the state derivatives. The simplest numerical method for calculating a derivative is to take the difference between the function at two points, then divide by the distance between the two points. The flowchart in Figure 4.3 shows this process used to calculate a Jacobian numerically. In this numerical evaluation, either the program must estimate increments for the independent variables, or the user must supply increments.

ALGORITHM

BACKWARD EULER INTEGRATION

The backward Euler integration algorithm can be used to solve sets of simultaneous differential equations. It has the general property that it will not become unstable for large ratios of integration interval to time constants.

1. Assume a starting value for the state vector, X_0, is given.
2. Solve the equation (possibly nonlinear) for $X(t + DT)$:

$$X(t + DT) = X(t) + f(X(t + DT), U(t + DT), t + DT).$$

3. A one-step method that can give good results for the solution for $X(t + DT)$ is

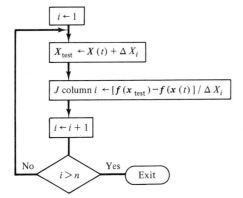

Figure 4.3 A flowchart for computing the Jacobian of $f(x(t), u(t), t)$.

$$\mathbf{X}(t+DT) = \mathbf{X}(t) + [I - DT\,J[\mathbf{X}(t), \mathbf{U}(t), t]^{-1}]\,DT\,\mathbf{f}(\mathbf{X}(t), \mathbf{U}(t), t)$$

with

$$J[\mathbf{X}(t), \mathbf{U}(t), t] = \text{Jacobian of } \mathbf{f}(\mathbf{X}(t), \mathbf{U}(t), t) = \left\{\frac{\partial f_i}{\partial x_j}\right\}$$

Depending upon the method chosen for solving the equation for $\mathbf{X}(t + DT)$, we actually have a family of algorithms that can be called "backward Euler" algorithms. Chua and Lin [6] point out that the Newton-Raphson iteration we have presented in step 3 permits accurate computation at larger integration intervals than other methods of solving the implicit, nonlinear expression for $\mathbf{X}(t + DT)$ that comes out of Step 2. For a complete discussion, see Chua and Lin.

PROBLEM 4.5

The block diagram of a phase-locked loop is shown. The phase detector has a nonlinear, sinusoidal characteristic. (There are several good accessible articles on phase-locked loops. The best is by Grebene [10]. Evaluate the Jacobian of this system analytically and compare analytical results with numerical results obtained using increments of 0.001 in each state in the Jacobian evaluation.

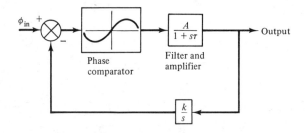

The Jacobian for this system cannot always be inverted. Under what conditions is this matrix inversion not possible? What are the implications of not being able to invert the Jacobian? Does this mean anything physically?

4.5 OTHER IMPLICIT ALGORITHMS

The backward Euler algorithm is perhaps the simplest possible implicit integration algorithm. However, there are other more complex and higher-order algorithms that can be useful. In this section we examine some of those algorithms and explore a family of related algorithms.

To begin consideration of other implicit algorithms we first look at a fairly simple one, the trapezoidal method. In this method, the expression of $\mathbf{X}(t + DT)$ is given by

$$\mathbf{X}(t+DT) = \mathbf{X}(t) + \frac{DT}{2}\mathbf{F}(\mathbf{X}(t), \mathbf{U}(t), t)$$

$$+ \frac{DT}{2}\mathbf{F}(\mathbf{X}(t+DT), \mathbf{U}(t+DT), t+DT) \qquad (4.5.1)$$

This expression uses the derivative at time t and at $t + DT$ and just averages the two to get an "effective" value of the derivative to use in computing the increment in $\mathbf{X}(t)$. This method might be expected to be more accurate than either the forward or backward Euler algorithms. Figure 4.4 suggests strongly that this is the case when we interpret the increment in $\mathbf{X}_i(t)$ as the area under the appropriate line.

Although this modified trapezoidal method looks simple, it is still an implicit algorithm. The unknown quantity we want to solve for, $\mathbf{X}(t + DT)$, still appears within the derivative function $\mathbf{F}$, on the right-hand side of the algorithm equation. Solving for $\mathbf{X}(t + DT)$ must again be done—using, for example, a Newton algorithm.

Solving for $\mathbf{X}(t + DT)$ in the expression for the trapezoidal algorithm, after linearizing around $\mathbf{X}(t)$ with the Jacobian, we have

$$\mathbf{X}(t+DT) = \mathbf{X}(t) + \frac{DT}{2}[\mathbf{F}(\mathbf{X}(t), t) + \mathbf{F}(\mathbf{X}(t+DT), t+DT)]$$

$$= \mathbf{X}(t) + \frac{DT}{2}\mathbf{F}(\mathbf{X}(t), t)$$

$$+ \frac{DT}{2}[\mathbf{F}(\mathbf{X}(t), t) + J(\mathbf{X}(t), t)\{\mathbf{X}(t+DT) - \mathbf{X}(t)\}] \qquad (4.5.2)$$

So,

$$\left[I - \frac{DT}{2}J(\mathbf{X}(t), t)\right]\mathbf{X}(t+DT) = \mathbf{X}(t) + DT\,\mathbf{F}(\mathbf{X}(t), t) - \frac{DT}{2}J(\mathbf{X}(t), t)\mathbf{X}(t) \qquad (4.5.3)$$

$$\mathbf{X}(t+DT) = \mathbf{X}(t) + \left[I - \frac{DT}{2}J(\mathbf{X}(t), t)\right]^{-1}DT\,\mathbf{F}(\mathbf{X}(t), t) \qquad (4.5.4)$$

Implementation of both the backward Euler and the modified trapezoidal algorithms is presented in the Appendixes. Both use the same computational method for the Jacobian, and there is very little difference in the implementation of the two algorithms. There is, however, some difference in the stability properties of the two algorithms.

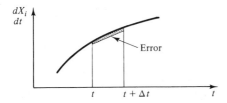

Figure 4.4 The trapezoidal algorithm.

If we examine the Euler, modified trapezoidal, and backward Euler algorithms, all applied to the first-order system

$$\frac{dx}{dt} = -ax \qquad (4.5.6)$$

we can compute expressions for $x(t + DT)$ for each of these algorithms. The results are given in Eq. (4.5.7).

Euler:

$$x(t + DT) = (1 - a\,DT)x(t) \qquad (4.5.7a)$$

Modified trapezoidal:

$$x(t + DT) = \frac{1 - a\,DT/2}{1 + a\,DT/2}x(t) \qquad (4.5.7b)$$

Backward Euler:

$$x(t + DT) = \frac{x(t)}{1 + a\,DT} \qquad (4.5.7c)$$

If a and DT are both positive numbers, then there is no way that the computed values for $x(t)$ can grow without bound when using the backward Euler. On the other hand, the Euler method has the possibility of becoming unstable if DT is chosen large enough. The modified trapezoidal method is an interesting "halfway" case. Clearly, the modified trapezoidal method can produce an oscillatory solution if $(1 - a\,DT/2)$ is negative. Nevertheless, if a and DT are both positive, then

$$\frac{1 - (a\,DT/2)}{1 + (a\,DT/2)}$$

will always have a magnitude less than one, and at least the computed solution will not grow without bound. The possibility of oscillation at larger integration intervals is the price paid for more accuracy at smaller integration intervals. Since everything really depends upon the ratio of integration interval to time constant, a stiff system could have oscillations in one mode while being very accurate in the "slower" mode. However, the most important point to be observed here is that a rapidly decaying mode will produce slowly decaying oscillations whenever the trapezoidal algorithm is used and the integration interval is much larger than the time constant of the rapidly decaying mode.

Finally, all of these algorithms can be given a circuit interpretation; that is, we can construct companion models for each of these algorithms. See Miniproject 3 at the end of this chapter.

4.6 GEAR'S ALGORITHMS

The problem of calculating the response of stiff systems has received a great deal of research attention. C. W. Gear has developed a set of implicit methods that have excellent stability properties [6, 11]. In this text we will not do an elaborate derivation of these methods, but we will present them and discuss some of their properties. We must also note that this is a very important set of algorithms that is becoming much more widely used. (ACSL permits the user to choose Gear's algorithms, and the *ACSL User Guide/Reference Manual* [8] has a good commentary on the relative merits of this and other algorithms.)

Gear's algorithms are all multistep; that is, they require knowledge of previously computed state values (before the present time, t) in order to compute the state, $X(t + DT)$. The single exception is the first-order Gear algorithm, which is our old friend, the backward Euler algorithm. Several higher-order Gear algorithms are listed in Table 4.3.

A sample derivation of one of the Gear algorithms is in order here. Consider the second-order Gear algorithm. All of Gear's algorithms involve the derivative of the state to be computed and only that derivative. (That helps ensure the stability of the algorithm. See Chua and Lin [6], pp. 524–529.) These algorithms also use past computed values of the states.

Now, imagine a situation in which the true solution is a polynomial. For the second-order Gear algorithm we will use a second-order polynomial. Gear's second-order algorithm should provide an exact solution, and we can determine the conditions under which that will occur. Let the polynomial be

$$X(t) = \alpha_0 + \alpha_1 t + \alpha_2 t^2 \tag{4.6.1}$$

If the algorithm is

$$X(t+DT) = a_0 X(t) + a_1 X(t-DT) + b_{-1} DT \, \dot{X}(t+DT) \tag{4.6.2}$$

then we can substitute the polynomial into the algorithm, getting

$$\begin{aligned}
\alpha_0 + \alpha_1(n+1) \, DT + &\alpha_2(n+1)^2 (DT)^2 \\
= a_0[&\alpha_0 + \alpha_1 n \, DT + \alpha_2 n^2 \, (DT)^2] \\
+ a_1[&\alpha_0 + \alpha_1(n-1) DT + \alpha_2(n-1)^2 DT^2] \\
+ b_{-1} \, DT[&\alpha_1 + 2\alpha_2(n+1) DT]
\end{aligned} \tag{4.6.3}$$

This must work for any such polynomial, so these expressions must hold for any polynomial coefficients. Look at each coefficient term separately and we find

$$\alpha_0: \qquad\qquad 1 = a_0 + a_1 \tag{4.6.4a}$$

$$\alpha_1: \qquad n+1 = na_0 + (n-1)a_1 + b_{-1} \tag{4.6.4b}$$

$$\alpha_2: \quad (n+1)^2 = n^2 a_0 + (n-1)^2 a_1 + 2(n+1)b_{-1} \tag{4.6.4c}$$

TABLE 4.3 GEAR'S ALGORITHMS

②	$X(t + DT) = \frac{4}{3}X(t) - \frac{1}{3}X(t - DT) + \frac{2}{3} DT \, \dot{X}(t + DT)$
③	$X(t + DT) = \frac{18}{11}X(t) - \frac{9}{11}X(t - DT) + \frac{2}{11}X(t - 2 \, DT) + \frac{6}{11} DT \, \dot{X}(t + DT)$
④	$X(t + DT) = \frac{48}{25}X(t) - \frac{36}{25}X(t - DT) + \frac{16}{25}X(t - 2 \, DT) - \frac{3}{25}X(t - 3 \, DT)$ $+ \frac{12}{25} DT \, \dot{X}(t + DT)$
⑤	$X(t + DT) = \frac{300}{137}X(t) - \frac{300}{137}X(t - DT) + \frac{200}{137}X(t - 2 \, DT)$ $- \frac{75}{137}X(t - 3 \, DT) + \frac{12}{137}X(t - 4 \, DT) + \frac{60}{137} DT \, \dot{X}(t + DT)$

Some algebraic manipulation yields three equations for the coefficients in the algorithm:

$$\left.\begin{array}{l} a_0 = 4/3 \\ a_1 = -1/3 \\ b_{-1} = 2/3 \end{array}\right\} \quad X(t+DT) = \tfrac{4}{3}X(t) - \tfrac{1}{3}X(t-DT) + \tfrac{2}{3}\dot{X}(t+DT) \quad (4.6.5)$$

The final solution is just Gear's second-order algorithm. Again, a FORTRAN implementation of the algorithm is given in the Appendixes.

4.7 AN EXTENDED EXAMPLE:
OPERATIONAL-AMPLIFIER CIRCUITS

Stiff systems are very frequently encountered in the world. A common stiff system is an operational amplifier. In the world of operational amplifiers the commonest species is the 741 model op amp.

If we want to compute transients in op-amp circuits, we need to have some sort of model for the amplifier itself. When these devices are first encountered the simplest model is one in which the device has infinite gain, infinite bandwidth, infinite input impedance, and zero output impedance. More realistic models can be constructed from manufacturer's specification sheets. However, there are many different models that can be constructed, depending upon the amount of complexity one can endure and the degree of precision that the problem solution demands.

The simplest op-amp model is a linear one constructed solely from the frequency response data given by a manufacturer's specification sheet. Figure 4.5 shows the open-loop gain and phase characteristics of a Signetics 741 op amp. Open-loop corner frequencies seem to be around 5 Hz and 2 MHz, upon examination of the open-loop gain characteristics. The phase characteristics support that conclusion, so the transfer function of the linear model of the op amp can be taken as something like

$$G_{\text{amp}}(s) = \frac{10^5}{(1 + s/30)(1 + s/1.2 \times 10^6)} \quad (4.7.1)$$

Here we have rounded the "2π" factor, since the numbers "5 Hz" and "2 MHz" were "guesstimates" in the first place. This model of the op amp has time constants of one-thirtieth of a second and less than a tenth of a microsecond. The ratio of time constants is almost a million, so the op amp represents a fairly stiff system.

Now, consider the op amp embedded within a circuit as in Figure 4.6. Imagine computing the step response of this circuit. To do this we can rearrange the transfer function to facilitate getting state equations. First, we note that the voltage to the inverting input, V_-, is given by

$$V_- = \left(\frac{R_f}{R_1 + R_f}\right)V_1 + \left(\frac{R_1}{R_1 + R_f}\right)V_{\text{out}} \quad (4.7.2)$$

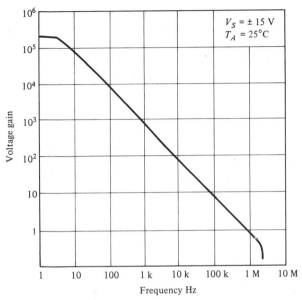

Figure 4.5 Open-loop frequency response for a 741 operational amplifier.

This expression can be combined with the transfer function information we already have relating V_{out} to V_- to yield the block diagram shown in Figure 4.7. That block diagram in turn can be translated into the state equations implemented in the FORTRAN program of Table 4.4.

Finally, this system is ready for simulation. Doing a simulation with a 1-μs integration interval produces instability using any of our explicit algorithms (Euler and second- and fourth-order Runge-Kuttas). Using the backward Euler (first-order Gear) produces the step response shown in Figure 4.8.

This kind of simulation shows how effective an implicit algorithm can be. In this particular case we obtain stable, accurate computations where explicit integration algorithms are unstable. However, for this particular linear system it is possible to devise special methods, taking advantage of the linearity of the system, to calculate transient response accurately and quickly. In Chapter 8 we will discuss those methods and develop techniques for calculation of transient response using matrix exponential methods. Still, however, the operational amplifier is not necessarily a linear device, and a more detailed model of the operational amplifier will not permit use of methods designed specifically for linear systems.

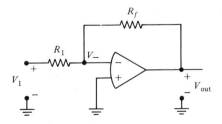

Figure 4.6 An operational-amplifier circuit.

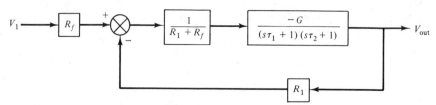

Figure 4.7 A block diagram model of the operational-amplifier circuit of Figure 4.6.

The most important nonlinear effects in an operational amplifier are output limiting and slew-rate limiting. In output limiting, the output voltage is limited to some specific value. For example, if the operational amplifier is powered by a dual 15-V supply (+15 V and −15 V), then the output voltage will never get above 13 or 14 V, or go below −13 or −14 V.

Trying to account for output limiting is not quite straightforward. One might attempt simply to add a "saturation" characteristic at the output of our block diagram model, as in Figure 4.9. The danger in this simple approach is that the output (which is state X_1 in the system) may be limited to 13 V (for example) but the input to the saturation nonlinearity may grow way beyond 13 V. If that happens, it may take a long wait until the input to the saturation element drops enough to permit the output to begin dropping also. This phenomenon is called "latch-up" and was present in many early integrated circuit operational amplifiers, but latch-up has been designed out of most amplifiers now available.

What we want to do is to provide some means for preventing the output (state X_1) from going above some saturation level, V_{max}. We can accomplish that by computing the derivative of the output, and checking (a) whether the derivative is positive, and (b) whether the output has just barely exceeded the saturation level, V_{max}. Whenever both conditions are satisfied, the state derivative is set to zero. A similar kind of situation exists around the negative saturation level, and a similar kind of computation can be done.

TABLE 4.4

```
C
C SUBROUTINE FOR LINEAR OP-AMP MODEL WITH TWO TIME-CONSTANTS
C
      SUBROUTINE XDOT (X,XD,T,N)
      DIMENSION X(20), XD(20)
C
      PARAMETER (GAIN=150000.,TAU1=.0318,TAU2=1.326E-8)
      PARAMETER (R1=1000., RF=100000.)
C
      VIN = 1.
      VMINUS = (RF*VIN + R1*X(1))/(R1+RF)
      XD(2) = -(GAIN*VMINUS + X(2))/TAU2
      XD(1) = (X(2) - X(1))/TAU1
C
      RETURN
      END
```

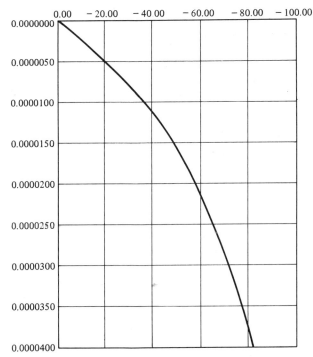

Figure 4.8 Simulation of the operational-amplifier circuit.

Slew rate must also be accounted for in a realistic operational-amplifier simulation. In this case we compute the rate we are "commanding" for the output, and if the commanded rate exceeds the slew rate, we set the commanded rate equal to the slew rate.

Figure 4.10 shows a complete nonlinear model that incorporates output limiting, slew-rate limiting, and both the high and low corner frequencies of the linear model.

Using our nonlinear model in a simulation produces some interesting results. With an integration interval of 5 μs, we find the following results.

1. For any explicit algorithm (Euler, second- and fourth-order Runge-Kuttas) the integration is unstable.
2. The backward Euler and the trapezoidal algorithms produce tolerable results. However, in both cases there are discrepancies from what we

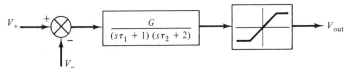

Figure 4.9 An operational-amplifier model with saturation.

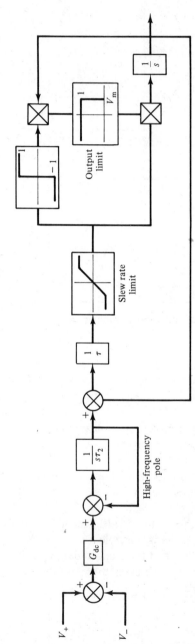

Figure 4.10 A refined operational-amplifier model.

might expect. In neither case does saturation occur at the 13.5 V level, but at some higher level instead.

3. The second-order Gear algorithm produces some very interesting results. The saturation in the system is not "enforced," and the results seem to show a very soft limiting.

The results here are a little surprising. However, the second-order Gear algorithm does tend to "remember" a trend. Thus, if there is a sharp discontinuity in the derivative, the second-order Gear algorithm tends to do some smoothing of the derivative calculation. That general behavior should also be found in higher-order Gear algorithms. For more enlightenment on this subject, we have included some problems in the miniprojects at the end of this chapter, ones with sharp nonlinearities.

4.8 CONCLUSIONS

In this chapter we have discussed various sorts of algorithms that are designed to give stable solutions of stiff dynamic systems. We are left with the feeling that there are still many loose ends. If you have to do simulations and you have a choice of integration algorithm, then you should do some experimenting, and you should give some thought to what is taking place in the calculation. This is still an area in which the user has to be wary.

In Chapter 8 we will discuss matrix exponential techniques for overcoming stiffness difficulties in linear system simulations. That is an area in which it is possible to overcome the stiffness problem with some confidence.

REFERENCES

1. A. Ralston and P. Rabinowitz, *A First Course in Numerical Analysis,* 2nd ed., McGraw-Hill, New York, 1978.
2. C. Runge, Uber die numerische Auflosung von Differentialgleichungen, *Math. Ann.* vol. 46, 1895, pp. 167–178.
3. W. Kutta, Beitrag zur naherungweisen Integration totaler Differentialgleichungen, *Z. Math. Phys.* vol. 46, 1901, pp. 435–453.
4. F. Scheid, *Theory and Problems of Numerical Analysis,* Schaum's Outline Series, McGraw-Hill, New York, 1968.
   References [1] and [4] are two of very few that actually present the Runge-Kutta derivations. Other sources that discuss these methods, but do not present a derivation of Runge-Kutta formulas, include books by Daniels and by Korn and Wait, listed in the Bibliography that follows.)
5. R. E. Scraton, *Basic Numerical Methods,* Edward Arnold, Ltd., London, 1984, pp. 66–70.
6. L. O. Chua and P-M Lin, *Computer Aided Analysis of Electronic Circuits: Algorithms and Computational Techniques,* Prentice-Hall, Englewood Cliffs, NJ, 1975.

7. R. L. Burden and J. D. Faires, *Numerical Analysis,* 3rd ed., PWS Publishers, Boston, 1985.

8. *Advanced Continuous Simulation Language (ACSL) User Guide/Reference Manual,* Mitchell & Gauthier Associates, Concord, MA, 1981.

9. N. H. McClamroch, *State Models of Dynamic Systems,* Springer-Verlag, New York, 1980. (A good source for other models and problems.)

10. A. B. Grebene, The monolithic phase-locked loop—A versatile building block, *IEEE Spectrum,* March 1971, pp. 38–49. (This is an excellent survey article on phase-locked loops. The article includes details of behavior and construction in a readable format.)

11. D. A. Calahan, *Computer-Aided Network Design,* rev. ed., McGraw-Hill, New York, 1972.

BIBLIOGRAPHY

R. W. Daniels, *An Introduction to Numerical Methods and Optimization Techniques,* North-Holland, New York, 1978.

G. A. Korn and J. V. Wait, *Digital Continuous System Simulation,* Prentice-Hall, Englewood Cliffs, NJ, 1978.

MINIPROJECTS

1. SPICE is probably the most widely used circuit analysis program at the time this book is being written. SPICE can be used to do transient analysis. Using SPICE compute the transient response of the circuit shown. Vary the component values to produce an even larger time-constant ratio. (As shown, the time-constant ratio is almost exactly a million to one, so the circuit equations are very stiff.)

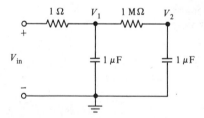

(a) From the way SPICE responds, would you expect the integration algorithm to be explicit or implicit?

(b) SPICE permits you to change algorithms. Using another algorithm redo this problem.

2. Complex poles in linear systems present different stability problems. For the Euler, second-order Runge-Kutta, and fourth-order Runge-Kutta experimentally determine pole locations in the complex plane for which the algorithms are stable for a unit integration interval. See Calahan's results for matrix exponential techniques and compare [10].

3. Several of the algorithms developed in this chapter can be used to generate companion models of capacitors and inductors. Develop companion models of capacitors and inductors for the modified trapezoidal algorithm. Compare the behavior of that algorithm and modeling technique with the backward Euler for the circuit of Figure 3.8.

4. The liquid-level simulation described in Problem 4.2 can be made into a liquid-level control problem. Referring to Problem 4.2, let the inflow rate be controlled by the difference between the desired height and the actual height, so that the inflow rate is given by:

$$\text{Inflow rate} = G(h_{\text{desired}} - h_{\text{actual}})$$

This will change the behavior of the system.

Assuming that the desired height is 0.25 m, determine how the steady-state height varies as a function of the gain, G, in the control law by doing simulations. Determine also how the "rise time" (the time to get to 90 percent of the steady-state value) is affected by the value of gain, G.

5. The interaction of field and armature currents in a DC motor produces interesting simulation problems. The state equations describing a DC motor are given in Reference [11] as

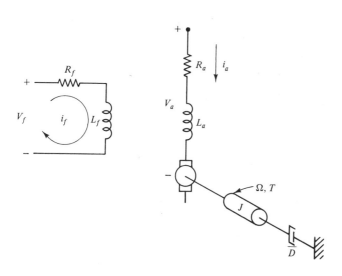

$$\frac{di_f}{dt} = -\left(\frac{R_f}{L_f}\right)i_f + \frac{V_f}{L_f} \quad \text{(Field current)}$$

$$\frac{di_a}{dt} = -\left(\frac{R_a}{L_a}\right)i_a - K_m K_f i_f \Omega \quad \text{(Armature current)}$$

$$\frac{d\Omega}{dt} = -\left(\frac{D}{J}\right)\Omega + K_m K_f \frac{i_a}{J} \quad \text{(Rotational speed)}$$

i_f = Field current (first state) $\qquad i_f(0) = 0$

i_a = Armature current (second state) $\qquad i_a(0) = 0$

Ω = Rotational velocity (third state) $\qquad \Omega(0) = 0$

$$\frac{L_f}{R_f} = \text{Field time constant} = 0.4 \text{ s}$$

$$\frac{L_a}{R_a} = \text{Armature time constant} = 0.05 \text{ s}$$

$$\frac{J}{D} = \text{Mechanical time constant} = 7.5 \text{ s}$$

$$L_f = 20 \text{ H}, \qquad K_m = 25, \qquad K_f = 0.02 \text{ Wb/A}, \qquad J = 1.5 \text{ kg} \cdot \text{m}^2$$

The large time-constant ratio makes calculation of the response of this motor difficult. The interaction between field current (first state) and rotational velocity (third state) in the state equation for armature current makes this a nonlinear system.

Calculate the response of this system when 200 V is suddenly applied to the armature and 100 V to the field coil. Find the peak torque and when it occurs. Use either Gear's first algorithm or the trapezoidal algorithm. Determine whether the peak torque and peak torque time of occurrence are sensitive to the integration interval used for the algorithm you choose.

Chapter 5

Optimization I

5.1 INTRODUCTION

We are all familiar with problems that involve finding optimum values of parameters. These kinds of problems can be posed in many different ways. For example, if we are designing the electric power system for an aircraft, we may need to design a transformer that weighs the least amount possible. Or, we may be concerned with maximizing the yield of a chemical process. Other examples include designing minimum-cost amplifiers, maximum-efficiency motors, or the most fuel-efficient automobile. All of these examples have one thing in common: They all are posed so that the solution to the problem involves adjusting parameters to minimize or maximize some single, scalar number.

The scalar that is "extremized" is usually referred to as a performance index, or simply as a PI. In some cases the performance index for a system will be obvious, and at other times it will not. Sometimes a performance index will be redefined as time goes on. For example, in the U.S. automobile market, the index used by the buyer is sometimes price, but it shifted toward fuel economy in the midst of the energy crisis. However, whatever the criterion used, it almost always is a single number, a scalar. Being able to reduce all of our concerns about a system to a single, quantifiable entity is almost a necessity if we are to do optimum designs using a digital computer.

We may not always be dealing with systems that have a well-defined performance index. For example, what if we want to design the best possible filter to filter noise from a signal that has frequency components up to 1 kHz? We really can't get very far unless we define what we mean by *best*, and "best" may not be defined for us by someone else.

We are going to be forced to assume that we can define some sort of per-

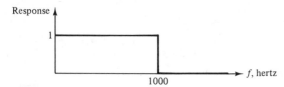

Figure 5.1 An ideal frequency response.

formance index in any system we work with and want to "optimize." Consider an example of how we might go about defining a performance index.

EXAMPLE 5.1

Consider the problem of designing a filter to reduce higher-frequency noise from an electric signal that has significant frequency components up to 1 kHz. Ideally, our filter will have a gain of unity up to 1 kHz and zero above 1 kHz. That ideal frequency response is shown in Figure 5.1. Any filter we build will have to deviate some what from this ideal, simply because this ideal is not "physically realizable." It has an impulse response that starts before zero, and any system that responds to an input before the input occurs simply can't ever be built. Still, we know that we can get passable approximations to this kind of frequency response, because we can always build a single-pole *RC* filter with a 3-dB point at 1 kHz.

The question that we hope to answer numerically is "How closely can the ideal be approximated?" To go any further, we have to define what we mean by "close." That definition will not be unique. However, one possible way of defining how far away we are from the ideal would be to compare the frequency response of the circuit to our ideal at a number of frequencies, and add up all the deviations at the different frequencies we examine. Table 5.1 show several possible ways of doing this.

We could take the measure of deviation as the absolute value of the difference in the responses (remembering we are dealing with complex numbers

TABLE 5.1

$$\sum_{i=1}^{N} |G_d(j\omega_i) - G(j\omega_i)|$$

$$\sum_{i=1}^{N} ||G_d(j\omega_i)| - |G(j\omega_i)||$$

$$\sum_{i=1}^{N} |G_d(j\omega_i) - G(j\omega_i)|^2$$

$$\sum_{i=1}^{N} [|G_d(j\omega_i)| - |G(j\omega_i)|]^2$$

where $G_d(j\omega)$ = Desired frequency response
         $G(j\omega)$ = Frequency response of circuit
         ω_i = Frequencies at which comparisons are made; $i = 1, \ldots, N$

when we deal with frequency responses), or we could take the square of that absolute value, or any other even power. Whatever we choose, we take that choice as our performance index for the filter, and in an optimization algorithm, that PI is what we would try to minimize.

5.2 SOME ANALYTICAL BACKGROUND

If we are trying to maximize or minimize some sort of performance index and do it analytically, we are really doing a calculus problem. However, since we may have many variables to "tweak," we will have a multidimensional calculus problem. In the example above, the filter might have a fixed form with several resistors and capacitors. Each resistor and capacitor would have to be chosen before the design was complete, and each resistor and capacitor would have an effect on the PI. Each parameter should be treated as a separate variable, and we will assume that we have a vector of parameters, and denote that vector or parameters by $\mathbf{X}$. We will also assume that our performance index is some function, $f(\mathbf{X})$, that we can plug the parameter vector into and calculate the PI.

The problem is thus redefined to one of finding a maximum or minimum of a function, $f(\mathbf{X})$, where $\mathbf{X}$ is a vector of parameter values. From calculus, we know that the condition for an extremum is that all the partial derivatives, with respect to all the vector components, be zero. Since the gradient is the vector of partial derivatives, what we must have is

$$\nabla(f(\mathbf{X})) = 0 \tag{5.2.1}$$

where ∇ is the gradient operator.

EXAMPLE 5.2

We will find the location of the minimum of the function

$$f(X_1, X_2) = f(\mathbf{X}) = (X_1 - 2)^2 + 2(X_2 - 4)^2 + 2$$

Form the gradient of the function, obtaining

$$\nabla f(\mathbf{X}) = \begin{bmatrix} 2(X_1 - 2) \\ 4(X_2 - 4) \end{bmatrix} = \begin{bmatrix} 0 \\ 0 \end{bmatrix}$$

Set the gradient to the zero vector to find the location of the minimum, then substitute those values into the function to find the minimum function value.

$$X_1 = 2$$
$$X_2 = 4$$
$$f(2, 4) = 2$$

EXAMPLE 5.3

The Rosenbrock function is well known as a test function for numerical optimization algorithms. The function is:

$$f_r(\mathbf{X}) = 100(X_2 - X_1^2)^2 + (1 - X_1)^2$$

Follow the same steps as in the previous example to find the location of the minimum, and the minimum function value.

$$\nabla f_r(\mathbf{X}) = \begin{bmatrix} 200(X_2 - X_1^2)(-2X_1) - 2(1 - X_1) \\ 200(X_2 - X_1^2) \end{bmatrix} = \begin{bmatrix} 0 \\ 0 \end{bmatrix}$$

$$X_2 = X_1^2 \qquad \text{(from second component)}$$
$$X_1 = 1 \qquad \text{(from first component)}$$

$X_1 = X_2 = 1$ gives a minimum function value:

$$f_r(1, 1) = 0$$

EXAMPLE 5.4

Find the minimum of the function, $f(\mathbf{X})$:

$$f(\mathbf{X}) = 11X_1^2 + 11X_2^2 - 18X_1X_2 + 18X_1 - 22X_2 + 21$$

Take the gradient of the function, obtaining

$$\nabla(f(\mathbf{X})) = \begin{bmatrix} 22X_1 - 18X_2 + 18 \\ -18X_1 + 22X_2 - 22 \end{bmatrix}$$

Now, set the gradient equal to zero, and the solution for the minimum is

$$X_1 = 0$$
$$X_2 = 1$$

PROBLEM 5.1

Find the location of the minimum and the minimum value of the function

$$f(\mathbf{X}) = 3X_1^2 - 18X_1 + 2X_2^2 - 4X_2 + 39$$

PROBLEM 5.2

Find the location of the minimum and the minimum value of the function

$$f(\mathbf{X}) = 100X_1^2 + 101X_2^2 + 200X_1X_2 - 200X_1 - 202X_2 + 99$$

The analytical requirement for a maximum or minimum is that the gradient must be zero at the extremum point. The gradient itself is an interesting entity. If we consider the change in a function as we move from a point $\mathbf{X}_0$ to another point $\mathbf{X}_1$, the change in the function will be given approximately by

$$f(\mathbf{X}_1) - f(\mathbf{X}_0) \simeq \nabla(f(\mathbf{X}_0))^\mathrm{T} \cdot (\mathbf{X}_1 - \mathbf{X}_0) \qquad (5.2.2)$$

Here we view the gradient, $\nabla f(\mathbf{X}_0)$, as an $n \times 1$ vector, and we form the dot product of the gradient and the increment in $\mathbf{X}$ by transposing the gradient (yielding a $1 \times n$ matrix) and then doing a matrix multiplication with the change in $\mathbf{X}$; that is, $\mathbf{X}_1 - \mathbf{X}_0$.

If we want to consider a minimization problem, and we are at some point $\mathbf{X}_0$, an interesting question is "What is the direction of the largest change of the PI function as we move about in the parameter space?" The way to get the most change is to line up the increment vector, $\mathbf{X}_1 - \mathbf{X}_0$, with the gradient. If we go in the direction of the gradient, we get the largest increase in $f(\mathbf{X})$, and if we go in the opposite direction, we get the largest decrease in $f(\mathbf{X})$. The reason for this behavior is that we are dealing with the dot product of two vectors, $\nabla f(\mathbf{X}))$ and $\mathbf{X}_1 - \mathbf{X}_0$. If both vectors are of fixed size, then the largest dot product occurs when the vectors are lined up or when they are $180°$ different. In the next section, we will take advantage of this behavior to construct an algorithm that searches out an extremum.

5.3 A GRADIENT ALGORITHM

If the gradient points in the direction of maximum change and if we are trying to improve the estimate of the location of the optimum, one strategy would be to move a small amount in the direction of the gradient (to maximize) or in the direction opposite to the gradient (to minimize). That turns out to be a reasonable, and widely used, algorithm, and one that we will now examine in more detail.

Assume that we have a minimization problem. We have to decide how large a step to take. We know the direction to go is opposite to the gradient direction, but choosing the step size is a separate decision. One reasonable approach might be to make the size of the step proportional to the gradient. Then, if we are far away from the optimum location, where the gradient is probably large, we will take larger steps than when we are close to the optimum location.

The algorithm we will use is to adjust an estimate of the location of the optimum by adding an increment to the estimate with the increment proportional to the gradient. First define terms, then define the details of the algorithm. Let

$$\mathbf{X}_n = \text{the } n\text{th estimate of the location of}$$
$$\text{the optimum (minimum)} \qquad (5.3.1)$$
$$\nabla(f(\mathbf{X}_n)) = \text{the gradient of the function, } f(\mathbf{X}),$$
$$\text{evaluated at the point, } \mathbf{X}_n \qquad (5.3.2)$$
$$G = \text{``gain'' constant, a scalar} \qquad (5.3.3)$$

Then the algorithm is

$$\mathbf{X}_{n+1} = \mathbf{X}_n - G * \nabla(f(\mathbf{X}_n)) \qquad (5.3.4)$$

It will help some if we can visualize what we are doing when we use this algorithm. To help this visualization, imaging that we want to minimize a function of two variables. We can sketch contours of equal value in the X_1-X_2 plane, as shown in Figure 5.2.

The function shown in Figure 5.2 has a minimum somewhere around the point ($X_1 = 2$, $X_2 = 3$). Surrounding the (2, 3) point are contours of equal values of the function. The algorithm takes steps in a direction along the gradient, perpendicular to the contours of equal function value. The solid line shows a sequence of steps for some gain value. The dashed line shows a sequence of steps using a higher value of gain G.

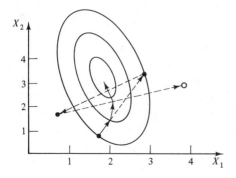

Figure 5.2 The gradient algorithm.

In the next two sections we will tackle two problems that arise. The first problem is that in many cases the gradient is not directly available and thus must be computed numerically. The second problem is the one already alluded to in Figure 5.2, that is, the problem of possible instability in the algorithm. If we view the performance index as a "hill" in the X_1-X_2 plane, we can see that there will be convergence problems if the hill is a "skinny ridge." Then the possibility of oscillations increases, and the skinnier the ridge, the worse the problem. We will look at this problem in more detail later. First, we will examine the problem of computing the gradient numerically.

ALGORITHM

METHOD OF STEEPEST DESCENT (GRADIENT ALGORITHM)

The gradient algorithm, often known as the "method of steepest descent," can be used to find the minimum (or maximum) of a function of N variables. The method searches in the N-dimensional "parameter space" by always moving in the direction opposite to the gradient of the function— that is, moving in the direction of steepest descent. The method iterates that step either until sufficient improvement in the function has been achieved or until some limit on number of iterations is reached. Individual steps in the algorithm follow the rule

$$\mathbf{X}_{k+1} = \mathbf{X}_k - G \, \nabla f(\mathbf{X}_k)$$

The user must specify some gain, G. The gradient can be computed either analytically or numerically, depending upon the individual problem being solved.

5.4 NUMERICAL COMPUTATION OF THE GRADIENT AND IMPLEMENTATION OF THE ALGORITHM

The partial derivatives in the gradient are really limits of differences, and can be approximated using the methods discussed in Chapter 2 for computing derivatives.

$$\nabla(f(\mathbf{X}))[j] = \frac{(f(\mathbf{X} + \mathbf{DX}_j) - f(\mathbf{X} - \mathbf{DX}_j))}{2\mathbf{DX}_j}$$

= Approximation to jth component of the gradient at X (5.41)

We would have to form N such differences to compute all N components of the gradient. We would have to make some sort of a priori choice of increments for each element of the parameter vector. Each element might be a different kind of variable—resistors, capacitors, gains, or whatever—and increments need not be the same numerical value.

Table 5.2 shows a typical program to compute a gradient numerically. In this program, $\mathbf{X}$ is the parameter vector, G is the computed value of the gradient, and STEP is the vector of step sizes used in incrementing $\mathbf{X}$ in the computation.

PROBLEM 5.3

Rewrite Subroutine GRAD to incorporate symmetric differences in computing each derivative component of the gradient.

Test your revised program on the function of Example 5.2, computing the gradient at (0, 0) and (1, 1).

In any computation of this sort, accuracy will fluctuate with the step size. If the step size is too large, then the function may not be "smooth" enough,

TABLE 5.2

```
      SUBROUTINE GRAD (X,G,STEP,N)
C
C GRAD EVALUATES THE GRADIENT OF A FUNCTION, FOFX
C
C    X = POINT AT WHICH THE GRADIENT IS EVALUATED
C        X IS THE PARAMETER VECTOR.
C
C    STEP = INCREMENTS IN X-COMPONENTS
C           WHEN EVALUATING THE GRADIENT.
C
C    G = COMPUTED VALUE OF THE GRADIENT.
C
C    N = NUMBER OF COMPONENTS IN X.
C
      DIMENSION X(10), G(10), STEP(10)
      CALL FOFX (X,N,F)
      DO 100 I = 1,N
      X(I) = X(I) + STEP(I)
      CALL FOFX (X,N,FDEL)
      G(I) = (FDEL - F)/STEP(I)
      X(I) = X(I) - STEP(I)
  100 CONTINUE
      RETURN
      END
```

fluctuating too much for this approach to evaluate the gradient accurately. On the other hand, if the increment is too small, then the computed values at **X** and **X** + **DX**$_j$, may be so close that roundoff effects make the difference meaningless. In any computation of this sort, the user of the algorithm must take care that the increments are chosen to give usable values for the gradient. It may be that only experience will let the user choose increments well, but the choice can usually be made after some experience with a particular kind of problem.

EXAMPLE 5.5

If we consider the problem of numerically evaluating the gradient of the function we used in the last example, we can get some feel for how this numerical process works. Let us imagine we are at the point (4.0, 5.0) in the (X_1, X_2) space. Assume further that we have arbitrarily decided to use increments of 0.01 and 0.02 for X_1 and X_2, respectively. Then, to compute the gradient, we would find

$$f(\mathbf{X}) = (X_1 - 2)^2 + 2(X_2 - 4)^2 + 2$$
$$f(4.0, 5.0) = 8.0$$
$$f(4.01, 5.0) = 8.0401$$
$$f(4.0, 5.02) = 8.0808$$

where $f(\mathbf{X}) = (X_1 - 2)^2 + 2(X_2 - 4)^2 + 2$.

So, we compute the gradient as

$$\nabla(f(4.0, 5.0)) = \begin{bmatrix} 4.01 \\ 4.040 \end{bmatrix}$$

which is close to the correct value of

$$\begin{bmatrix} 4 \\ 4 \end{bmatrix}$$

5.5 STABILITY OF THE ALGORITHM

Consider the simplest possible optimization problem, a quadratic function of a single variable. A general quadratic function is

$$f(x) = ax^2 + bx + c \qquad (5.5.1)$$

This is a very simple function to optimize numerically, but it does exhibit some of the peculiarities and problems that are found in more complex problems, so it is worth considering. The gradient in this case is just the derivative with respect to the scalar variable, x. Taking the derivative, we have

$$\frac{df(x)}{dx} = 2ax + b \qquad (5.5.2)$$

Using these expressions in our gradient algorithm, we have

$$x_{k+1} = x_k - g(2ax_k + b)$$
$$= (1 - 2ga)x_k - gb \qquad (5.5.3)$$

Now, if this iterative procedure ever reaches a steady state, the steady state will be when $x_{k+1} = x_k$, or when

$$2gax_k = -gb \qquad (5.5.4)$$

or

$$x_k = \frac{-b}{2a} \qquad (5.5.5)$$

We are not guaranteed that this iterative procedure will ever reach this steady state. The iteration may well be unstable. However, at least the steady state, if we reach it, is the correct solution.

We can obtain the theoretical solution to the difference equation formed by the algorithm. That solution is

$$x_k = (1 - 2ga)^k x_0 - \frac{b}{2a} \qquad (5.5.6)$$

The term multiplying the initial guess, x_0, must die out if the theoretical solution is to approach the correct solution, $-b/2a$. However, if the absolute value of $(1 - 2ga)$ is larger than unity, then the x_0 term will grow without bound as we iterate, and the computed solution will never converge to the true solution. We can guarantee convergence in this case if we require

$$g < \frac{1}{2a} \qquad (5.5.7)$$

If the gain is too large, then the algorithm becomes unstable for this function.

In general, we will not know the details of the function we are trying to optimize. If we did, we might be able to generate an analytic solution. Since we don't know those details, we may be in the position of guessing a value of gain. However, if we encounter instability, we should conclude that our gradient gain is too large, and we should explore smaller values. A word of caution is in order. It sometimes is the case that gain has to be reduced many orders of magnitude to obtain stability, and it never seems obvious ahead of time what value of gain to use. Looking at the numerical value of the gradient can help immensely when trying to choose gain values.

EXAMPLE 5.6

In Example 5.5 we considered a function

$$f(\mathbf{X}) = (X_1 - 2)^2 + 2(X_2 - 4)^2 + 2$$

The gradient for the function was

$$\begin{bmatrix} 1 \\ 4 \end{bmatrix}$$

Obviously, if we used a gain of 10^{-10}, we would take such small steps that it would take forever to accumulate any change in the location of X_1 and X_2. (Roundoff may make it take forever literally!) Conversely, with a gain of 1000, we would be making changes that would be far too large. Clearly, there is a middle ground, and although there is a lot of space in the middle ground, it does help to examine the gradient to see where that middle lies. That will at least prevent us from using gain values that could never work.

5.6 PROGRAMS FOR DOING GRADIENT OPTIMIZATIONS

In Section 5.4 we presented a program for computing the gradient of an arbitrary function. Now we would like to take that gradient computing program and use it to find the optima of some simple functions. In a later section we will define a more complex circuit optimization problem and use the routines we are now going to develop to compute solutions for those kinds of problems. However, we will consider a few simpler problems first, just to get a feel for the numerical techniques.

In order to use the gradient method, a few numerical preliminaries have to be taken care of. These are as follows:

1. The starting values of the parameters must be defined.
2. The step sizes for computing the gradient must also be defined.
3. The gain in the gradient algorithm needs to be defined.

A main program for entering this data is given in Table 5.3.

The main program in Table 5.3 takes care of all the data input that is necessary for doing a gradient optimization. In addition, this main program also does the gradient optimization, printing every result as the search proceeds. If those results are not desired, the print can easily be eliminated from the program. However, a student just doing a few first problems may find it more educational to follow the search process once or twice, so the program as given prints all data output.

Besides the main program, we need to write a program that actually gives the function we are going to minimize numerically. In the routines above, that function is a subroutine called FOFX, a mnemonic for "F of X." Anyone using this system will have to write individual programs for each particular minimization problem, but the routines have been modularized to separate the algorithm implementation from the numerical definition of the function being minimized.

EXAMPLE 5.7

Let us write an FOFX subroutine and use it to minimize the function

$$F(\mathbf{X}) = X_1^2 + 10(X_2 - 3)^2 + 17$$

The subroutine for this function is given in Table 5.4.

TABLE 5.3

```
C
C
C THIS IS A MAINLINE TO DO GRADIENT OPTIMIZATION
C
C THIS PROGRAM ASSUMES THAT THE USER HAS GENERATED
C A FUNCTION PROGRAM, FOFX, WHICH RETURNS THE
C VALUE OF THE FUNCTION TO BE MINIMIZED.
C
C PLEASE NOTE, WE ARE ASSUMING A MINIMIZATION PROBLEM.
C
C
      DIMENSION X(10), DELX(10)
      DIMENSION G(10)
C
C X IS THE PARAMETER VECTOR.
C G IS THE COMPUTED GRADIENT VALUE.
C N IS THE SIZE OF X AND G.
C DELX IS IHE STCP GIZE FOR COMPUTING THE GRADIENT.
C
C READ IN THE INITIAL GUESS FOR THE OPTIMUM LOCATION.
C
      WRITE (6,*) ' INPUT THE NUMBER OF PARAMETERS.'
      READ (5,*) N
      WRITE (6,*) ' INPUT THE STARTING VALUES OF PARAMETERS.'
      READ (5,*) (X(I), I=1,N)
      WRITE (6,*) ' INPUT THE STEPS FOR EACH PARAMETER'
      WRITE (6,*) ' TO BE USED IN GRADIENT EVALUATION.'
      READ (5,*) (DELX(I), I=1,N)
      WRITE (6,*) ' HOW MANY STEPS DO YOU WANT TO DO'
      WRITE (6,*) ' IN YOUR OPTIMIZATION RUN?'
      READ (5,*) NSTEPS
      WRITE (6,*) ' WHAT GAIN VALUE DO YOU WANT TO USE?'
      READ (5,*) GAIN
      WRITE (6,*) ' PARAMETER VALUES AND FUNCTION VALUE'
C
C PRINT OUT EVERY STEP WITH THIS VERSION
C
      DO 100 I=1,NSTEPS
      CALL GRAD (X,G,DELX,N)
      DO 20 K=1,N
      X(K) = X(K) - GAIN*G(K)
   20 CONTINUE
      CALL FOFX (X,N,F)
      WRITE (6,*) (X(J), J=1,N), F
  100 CONTINUE
      CALL EXIT
      END
```

TABLE 5.4

```
      SUBROUTINE FOFX(X,N,F)
      DIMENSION X(10)
C
C THIS IS AN EXAMPLE SUBROUTINE
C
      F= (X(1)**2) + 10.*((X(2)-3.)**2) + 17.
      RETURN
      END
```

This function has a minimum at $X_1 = 0$ and $X_2 = 3$. Table 5.5 gives the results obtained by running this FOFX using a starting point of (10, 10) and a gain of 0.025 in the gradient algorithm. These results show a distinct difference in the rate of convergence of X_1 and X_2. In one case, X_2, convergence to the optimum value, 3, is obtained relatively quickly, but for X_1, convergence to 0 is relatively slow.

TABLE 5.5

```
INPUT THE NUMBER OF PARAMETERS.
?2,

INPUT THE STARTING VALUES OF PARAMETERS.
?10., 10.,

HOW MANY STEPS DO YOU WANT TO DO
IN YOUR OPTIMIZATION RUN?
?25,

WHAT GAIN VALUE DO YOU WANT TO USE?
?.025,

PARAMETER VALUES AND FUNCTION VALUE
9.504089       6.513367       230.7652
9.029159       4.759083       129.4694
8.578071       3.881225       98.34888
8.149871       3.441104       85.36613
7.743129       3.221282       77.44571
7.355937       3.111133       71.23332
6.988296       3.056296       65.86797
6.639251       3.028640       61.08785
6.307372       3.014335       56.78499
5.992182       3.007659       52.90684
5.692729       3.004321       49.40734
5.408057       3.001936       46.24711
5.137690       3.000983       43.39587
4.881151       3.000983       40.82565
4.637011       3.000506       38.50187
4.405268       3.000982       36.40639
4.185446       3.000982       34.51796
3.976114       3.000505       32.80948
3.777750       3.000505       31.27139
3.589399       3.000505       29.88379
3.410108       3.000505       28.62884
3.239877       3.000505       27.49681
3.078229       3.000505       26.47550
2.924687       3.000504       25.55380
2.778775       3.000504       24.72159
```

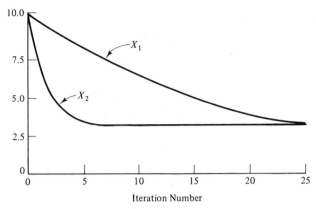

Figure 5.3 X_1 (upper curve) and X_2 (lower curve) as functions of the number of iterations.

    The reason for the difference in the rates of convergence is simply that the contours of equal function value form a relatively "skinny" ellipse in the X_1-X_2 plane.

    For this function, the two parameters, X_1 and X_2 are "uncoupled," and the gradient adjustment algorithm is really two one-dimensional processes:

$$X_{1(k+1)} = X_{1k} - 2G(X_{1k} - 0) \tag{5.6.1a}$$

and
$$X_{2(k+1)} = X_{2k} - 20G(X_{2k} - 3) \tag{5.6.1b}$$

Each separate iterative process has its own rate of convergence. In this case, one process could converge while the other one diverges.

    It is somewhat more informative to plot this data. Figure 5.3 is such a plot, with X_1 and X_2 plotted against iteration number (from 1 to 25). The variable X_1 takes considerably longer to reach its equilibrium value compared to the variable X_2, which reaches its equilibrium value in just a few iterations. The difference in convergence rates in the two variables is obvious and somewhat striking.

    While the independent variables X_1 and X_2 are converging to their ultimate values, the performance index is also converging to its ultimate value, and it is worthwhile looking at how that convergence takes place. Figure 5.4 shows a plot of performance index against iteration number. Comparing the convergence of the performance index to that of the variables, it looks as though the performance index converges quickly, like X_2 does, and not as slowly as X_1 converges.

5.7 QUADRATIC REPRESENTATIONS OF PERFORMANCE INDICES

In the preceding section we encountered an example of a function that exhibited different rates of convergence to the maximum point in the two parameters. In

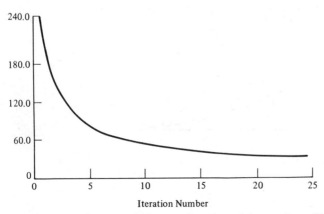

Figure 5.4 Performance index as a function of the number of iterations.

the next section we will examine that phenomenon in some more detail. First, we need to consider how we can represent a general quadratic function.

Any performance index will probably be smooth near an extremum point. If the performance index function is smooth enough, it can be approximated by a truncated multidimensional Taylor series, one in which the coefficients are various partial derivatives with respect to the parameters. Let us assume that is the situation we have. If we are close enough to the optimum point, we will be able to represent the function with constants, plus linear and quadratic terms.

If we limit ourselves to quadratic functions, then, we can begin to think about how to represent these sorts of multidimensional quadratic functions. If we consider a two-dimensional parameter vector, then the most general quadratic function we can write is

$$F(X_1, X_2) = A_{11}X_1^2 + A_{12}X_1X_2 + A_{22}X_2^2 + B_1X_1 + B_2X_2 + C \qquad (5.7.1)$$

This function is composed of a constant C, linear terms B_1 and B_2, and product terms, A_{11}, A_{12}, A_{22} (which involve all possible cross-products and self-products among the parameters X_1 and X_2). With more elements in the parameter vector, we would have to include a larger number of essentially similar terms.

A quadratic expression involving quadratic, linear, and constant terms can be expressed using matrix and vector representations. Let us assume that we have a parameter vector, **X**, defined as

$$\mathbf{X} = \begin{bmatrix} X_1 \\ \vdots \\ X_N \end{bmatrix} \qquad (5.7.2)$$

Then a combination of linear terms in the elements of the parameter vector can be generated by taking the inner (dot) product between the parameter vector and a constant vector:

$$\text{Linear terms} = \mathbf{B}^\mathsf{T}\mathbf{X} \qquad (5.7.3)$$

Here, $\mathbf{B}^T$ is the transpose of the vector, $\mathbf{B}$. Similarly, quadratic terms can be generated using pre- and postmultiplication of a constant matrix by the parameter vector; that is,

$$\text{Quadratic terms} = \mathbf{X}^T A \mathbf{X} \qquad (5.7.4)$$

Thus we can represent a general quadratic function with the form

$$f_{\text{quad}}(\mathbf{X}) = (\mathbf{X}^T A \mathbf{X}) + (\mathbf{B}^T \mathbf{X}) + C \qquad (5.7.5)$$

where

$$\begin{array}{ll}
\mathbf{X} \text{ is an } N \times 1 \text{ vector} & (5.7.6a) \\
A \text{ is an } N \times N \text{ matrix} & (5.7.6b) \\
\mathbf{B} \text{ is an } N \times 1 \text{ vector} & (5.7.6c) \\
C \text{ is a scalar} & (5.7.6d)
\end{array}$$

EXAMPLE 5.8

Represent the function

$$f_{\text{quad}}(X_1, X_2) = 21 * (X_1^2) - 38X_1 X_2 + 21X_2^2 - 82X_1 + 82X_2 + 81$$

Using the formulation above, we have

$$\text{Constant term} = 81$$
$$\text{Linear term} = -82X_1 + 82X_2$$
$$= [-82, \quad 82] * \begin{bmatrix} X_1 \\ X_2 \end{bmatrix}$$

$$\text{Quadratic term} = [X_1, \quad X_2] * \begin{bmatrix} 21 & -19 \\ -19 & 21 \end{bmatrix} * \begin{bmatrix} X_1 \\ X_2 \end{bmatrix}$$

so

$$A = \begin{bmatrix} 21 & -19 \\ -19 & 21 \end{bmatrix} \qquad \mathbf{B} = \begin{bmatrix} -82 \\ 82 \end{bmatrix} \qquad C = 81$$

If we have a quadratic model, then there are two natural questions that arise. The first question is: "Just where is the minimum value of the function?" The second question concerns the rate(s) of convergence in using the gradient optimization algorithm in this situation.

We will first find the location of the optimum point. If we consider taking the gradient of the quadratic function, we can get a general form for the result. Taking the terms one at a time, starting with the (simplest) constant term, we have zero for the gradient of the constant. The general linear term is

$$B_i X_i \qquad (5.7.7)$$

Taking the partial derivatives of this term will yield zero except when taking the derivative with respect to X_i. So this term will contribute zero to all components of the gradient, excepting the ith component. The ith component will be just

B_i, the ith component of the vector **B**, so this part of the gradient will just be the vector **B**.

The quadratic terms present more of a problem. Here the general term is

$$A_{ij}X_iX_j \tag{5.7.8}$$

When the two indices i and j are different and we take the derivative with respect to one of the parameter vector components—say, X_i—then we obtain a part of the ith component of the gradient. That component is

$$A_{ij}X_j \tag{5.7.9}$$

In the matrix A, there will be two terms that involve the same indices, i and j. We will *assume* that $A_{ij} = A_{ji}$. The X_iX_j term can be split any number of ways, but the total X_iX_j term is $(A_{ij} + A_{ji})X_iX_j$, and it is really irrelevant how that split is made, since everything gets added together in the end anyway. So assuming that $A_{ij} = A_{ji}$ causes no loss in generality.

The terms involving squares, $A_{ii}X_i^2$, require a different treatment. Here the contribution to the ith component of the gradient is simply $2A_{ii}X_i$.

Combining all the quadratic terms, we find

$$\nabla(\mathbf{X}^{\mathrm{T}}A\mathbf{X}) = 2A\mathbf{X} \tag{5.7.10}$$

and
$$\nabla(f_{\text{quad}}(\mathbf{X})) = \nabla((\mathbf{X}^{\mathrm{T}}A\mathbf{X}) + (\mathbf{B}^{\mathrm{T}}\mathbf{X}) + C) = 2A\mathbf{X} + \mathbf{B} \tag{5.7.11}$$

EXAMPLE 5.9

Find the gradient of the quadratic function used in the previous example. Here

$$A = \begin{bmatrix} 21 & -19 \\ -19 & 21 \end{bmatrix} \qquad \mathbf{B} = \begin{bmatrix} -82 \\ 82 \end{bmatrix} \qquad C = 81$$

Using the results just developed, we must have the gradient of the quadratic function:

$$\nabla(f_{\text{quad}}(\mathbf{X})) = 2A\mathbf{X} + \mathbf{B}$$
$$= \begin{bmatrix} 42 & -38 \\ -38 & 42 \end{bmatrix} * \begin{bmatrix} X_1 \\ X_2 \end{bmatrix} + \begin{bmatrix} -82 \\ 82 \end{bmatrix}$$

Once we have the gradient of a quadratic function, we can then determine where the optimum is located by setting the gradient to zero. Doing that, we obtain

$$\nabla(f_{\text{quad}}(\mathbf{X})) = 0$$
$$= 2A\mathbf{X} + \mathbf{B} \tag{5.7.12}$$

Then, solving for the optimum **X**, we find

$$\mathbf{X}_{\text{opt}} = -\tfrac{1}{2}A^{-1}\mathbf{B} \tag{5.7.13}$$

From this result we can conclude that we could determine the optimum location immediately if we could find A and **B** and if we could reliably compute the inverse of matrix A. In a later section, we will discuss some methods that gather information on A and B as a search proceeds and are able to improve on the efficiency of a straightforward gradient search.

PROBLEM 5.4

Find the quadratic matric representation of the function

$$f(\mathbf{X}) = 11X_1^2 + 11X_2^2 - 18X_1X_2 + 18X_1 - 22X_2 + 21$$

PROBLEM 5.5

Find the quadratic matric representation of the function

$$f(\mathbf{X}) = 100X_1^2 + 101X_2^2 + 200X_1X_2 - 200X_1 - 202X_2 + 99$$

5.8 STABILITY PROPERTIES OF THE MULTIDIMENSIONAL GRADIENT SEARCH ALGORITHM

With the vector matrix representation we now have for multidimensional quadratic functions, we can examine the stability properties of the algorithm when used for multidimensional searches.

If we assume we have a quadratic representation for our performance index, then we have

$$f_{\text{quad}}(\mathbf{X}) = \mathbf{X}^T A \mathbf{X} + \mathbf{B}^T \mathbf{X} + C \tag{5.8.1}$$

The gradient is given by

$$\nabla(f_{\text{quad}}(\mathbf{X})) = 2A\mathbf{X} + \mathbf{B} \tag{5.8.2}$$

Then, using this expression in the gradient search algorithm, we have:

$$\begin{aligned}
\mathbf{X}_{k+1} &= \mathbf{X}_k - G\,\nabla(f_{\text{quad}}(\mathbf{X}_k)) \\
&= \mathbf{X}_k - G(2A\mathbf{X}_k + \mathbf{B}) \\
&= (I - 2GA)\mathbf{X}_k - G\mathbf{B}
\end{aligned} \tag{5.8.3}$$

Now, to determine the transient performance we will use Z-transform methods. Take Z-transforms of both sides of the last expression for $\mathbf{X}_{k+1}$, remembering that we are dealing with vectors and matrices. We obtain

$$z\mathbf{X}_z - z\mathbf{X}_0 = (I - 2GA)\mathbf{X}(z) - G\mathbf{B}\left(\frac{z}{z-1}\right) \tag{5.8.4}$$

Here, we have assumed that the constant term, GB, is a step function. Now, we can solve for $X(z)$, getting the result

$$\mathbf{X}_z = -\left[(zI - (I - 2GA))^{-1}G\mathbf{B}\left(\frac{z}{z-1}\right)\right] \tag{5.8.5}$$

The poles in the function X_z are at $+1$ (from the step input) and wherever we find eigenvalues of the matrix $(I - 2GA)$. If those eigenvalues lie outside of the unit circle in the z plane, then our gradient algorithm will be unstable.

EXAMPLE 5.10

Let us compute what happens in the case of a function without any "cross terms." Consider the function

$$f(\mathbf{X}) = L_0 + L_1 X_1^2 + L_2 X_2^2$$

In this case the parameters in the general quadratic representation are

$$A = \begin{bmatrix} L_1 & 0 \\ 0 & L_2 \end{bmatrix} \qquad \mathbf{B} = \mathbf{0} \qquad C = L_0$$

The gradient algorithm takes the form:

$$\mathbf{X}_{k+1} = \begin{bmatrix} 1 - 2GL_1 & 0 \\ 0 & 1 - 2GL_2 \end{bmatrix} * \mathbf{X}_k$$

Then the eigenvalues that must remain inside the unit circle are

$$1 - 2GL_1 \quad \text{and} \quad 1 - 2GL_2$$

Since both of these eigenvalues must be inside the unit circle, we must have

$$G < \text{MIN} \left((1/L_1), (1/L_2) \right)$$

Now we will examine some specific values. Assume that $L_1 = 1$ and $L_2 = 100$. Then we must have:

$$G < \text{MIN} (1, 100) = 1$$

However, if we pick $G = 0.5$, to keep it well within the allowable limit, we would have

$$X_{1(k+1)} = [1 - 2(0.5)]X_{1k} = 0$$
$$X_{2(k+1)} = [1 - 2(0.5)(0.01)]X_{2k} = 0.99 X_{2k}$$

The price we pay for stability in the X_1 computation is very slow convergence in the X_2 computation.

The kind of behavior we find in this example is typical of the kind of problem we encounter. Although we will not always have parameters that do not interact through cross terms, we will frequently find that the price of stability will be slow convergence if the eigenvalues of $(I - 2GA)$ are widely different (that is, differences of one or more orders of magnitude).

PROBLEM 5.6

Find the vector matrix representation for the functions below. From the vector matrix model, determine the location of the minimum of the function, and the value of the minimum. Finally, find the minimum experimentally, using the programs given in earlier sections, writing the requisite FOFX subroutines.

1. $f_a(\mathbf{X}) = X_1^2 + 10X_2^2 - 2X_1 - 80X_2 + 5$
2. $f_b(\mathbf{X}) = 111X_1^2 + 111X_2^2 + 100X_3^2 + 20X_1X_2 + 2X_1X_3 + 2X_2X_3 - 82X_1 - 78X_2 + 161$

5.9 AN EXAMPLE: OPTIMIZATION OF THE FREQUENCY RESPONSE OF AN *RC* NETWORK

In this section we will examine the problem of getting the frequency response of an *RC* ladder network to conform as closely as possible to the ideal bandpass characteristic. We will consider the problem of defining a performance index for the network, one that measures the deviation of the network's actual performance from our ideal performance. Then we will examine how to implement that performance index in a FORTRAN subroutine. Finally, we will pose the problem of finding the optimum numerically, and examine the problems in trying to optimize the network numerically.

The network configuration we will use is shown in Figure 5.5. We will assume that the two resistors are variable and that the two capacitors are fixed. This will effectively allow us to vary the location of the two poles, or corner frequencies of the network. The DC gain is fixed at unity for this configuration, and the two pole locations are the only other items left to be determined in the second-order transfer function of the network, so varying the two resistors will suffice.

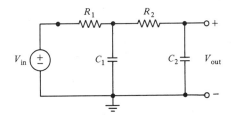

Figure 5.5 An *RC* network.

The essence of our problem is to define a performance index. We have to define some measure of how far away we are from the ideal frequency response.

Let us assume that we are trying to design a filter with a 1-kHz bandwidth. Then the ideal frequency response is the one previously given earlier in Figure 5.1. That ideal frequency response is shown in Figure 5.6, along with a superposed

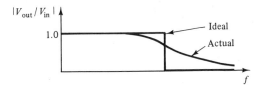

Figure 5.6 An ideal and an actual frequency response.

response of a real filter. We see that the real response deviates from the ideal response at all frequencies, with the largest deviation in the neighborhood of the passband-to-stopband transition. In order to get some measure of total deviation, let us assume that we take the difference between the actual and the ideal for some set of frequencies, and add up the deviations at all the frequencies in our set.

We will do these computations using computer routines. In this case we will break the computational task into two subtasks. First, we will write a routine that can compute the circuit's response at any frequency. Then we will write a routine that compares the circuit's response to the ideal response and does the computation of the performance index.

The FORTRAN routine RCFILTER, given in Table 5.6, is the routine that computes frequency response of the RC network. In RCFILTER, the capacitors are assumed to be 0.001 μF. Otherwise, this routine does a straightforward computation of the response of the network, using a previously derived formula for the transfer function of the network.

Next, we look at a routine that computes a sum of the squares of the

TABLE 5.6

```
      SUBROUTINE RCFILTER (X,FREQ,G)
C
C THIS SUBROUTINE CALCULATES THE FREQUENCY RESPONSE
C OF A NETWORK WITH THE TRANSFER FUNCTION:
C
C --------------------1---------------------------
C (S**2)*(R1*R2*C1*C2) + S*(R2*C2+R1*C1+R1*C2) + 1
C
      DIMENSION X(10)
      COMPLEX G
C
C R1 AND R2 ARE ASSUMED VARIABLE.
C
      R1 = X(1)
      R2 = X(2)
C
C C1 AND C2 ARE ASSUMED CONSTANT (.001 MICROFARADS).
C
      C1 = 1.E-9
      C2 = 1.E-9
C
      OMEGA = 6.2831853*FREQ
      GIMAG = OMEGA*((R1*C1)+(R2*C2)+(R1*C2))
      GREAL = 1. - (OMEGA**2)*(R1*R2*C1*C2)
      G = GREAL*(1.,0.) + GIMAG*(0.,1.)
      G = (1.,0.)/G
      RETURN
      END
```

differences between the ideal response (GD_I in the program) and the actual response (G_I). We use the square of the differences because that is a simple function that eliminates the possibility of a positive difference canceling out a negative difference. Since we want to use this routine with our gradient optimization algorithm, we use FOFX as the name of the routine, and we use arguments compatible with the gradient optimization routines. The routine FOFX is given in Table 5.7. In this FOFX routine we take the difference in magnitude of the ideal and actual response, then square and sum.

TABLE 5.7

```
      SUBROUTINE FOFX(X,N,F)
C
C THIS ROUTINE CALCULATES THE DIFFERENCE BETWEEN
C THE DESIRED FREQUENCY RESPONSE, GD, AND THE
C ACTUAL FREQUENCY RESPONSE, G.
C
      DIMENSION X(10)
      DIMENSION FREQ(10)
C
C FREQ HOLDS THE FREQUENCIES AT WHICH RESPONSE
C IS COMPARED
C
      COMPLEX GD(10)
      COMPLEX G
C
C GD HOLDS THE DESIRED RESPONSE AT FREQUENCIES
C STORED IN ZERO
C
C
C INITIALIZE THE DISTANCE
C
      F = 0.
      DO 40 I=1,10
      FREQ(I) = -100. + 200.*FLOAT(I)
C
      IF (FREQ(I) .LE. 1000.) GD(I)=1.
      IF (FREQ(I) .GE. 1000.) GD(I)=0.
C
      CALL RCFILTER(X,FREQ(I),G)
      DIFF = CABS(GD(I)) - CABS(G)
C
C COMPUTE THE DISTANCE FROM THE DESIRED RESPONSE.
C
      F = F + DIFF**2
   40 CONTINUE
      RETURN
      END
C
```

Before we can use this routine, we need to determine what values of resistances we should start from for R_1 and R_2. Since both capacitances are 0.001 μF, we can at least get an order-of-magnitude estimate by assuming that we have two uncoupled RC networks with the same time constant. Doing that, we would need to have the RC product equal to $1/2\pi(1000)$ to get corner frequencies at 1 kHz. We can use that to get a ballpark figure for R_1 and R_2, and we will therefore start with values of 160,000 Ω for each resistor.

Using these starting values, we get the results shown in Table 5.8. Note some peculiarities of those results. We need an extraordinarily high gain. Secondly, we find that we can have quite large changes in the resistance values, without correspondingly large changes in the performance index.

TABLE 5.8

```
?
 INPUT THE NUMBER OF PARAMETERS.

?
2,
 INPUT THE STARTING VALUES OF PARAMETERS.

?
160000.,160000.,
 INPUT THE STEPS FOR EACH PARAMETER
 TO BE USED IN GRADIENT EVALUATION.

?
10.,10.,
 HOW MANY STEPS DO YOU WANT TO DO
 IN YOUR OPTIMIZATION RUN?

?
50,
 WHAT GAIN VALUE DO YOU WANT TO USE?

?
5.E+9,
 PARAMETER VALUES AND FUNCTION VALUE
143251.1        153681.9        1.103144
129184.4        148675.1        1.064119
118694.0        145396.9        1.044049
111839.4        143727.9        1.036150
107786.3        143251.1        1.033612
105580.9        143608.7        1.032815
104329.2        144264.3        1.032455
103494.8        145158.4        1.032178
102839.1        146112.1        1.031929
102421.9        147065.7        1.031709
102064.2        148079.0        1.031491
101587.4        148973.1        1.031290
```

TABLE 5.8 (continued)

101170.2	149807.5	1.031111
100812.5	150642.0	1.030941
100454.9	151476.5	1.030777
100097.3	152310.9	1.030617
99739.63	153085.8	1.030472
99382.00	153860.6	1.030331
99024.37	154575.9	1.030203
98666.73	155291.1	1.030079
98368.71	156006.4	1.029963
98070.69	156721.6	1.029850
97772.66	157377.3	1.029748
97474.64	158032.9	1.029650
97236.22	158688.6	1.029558
96938.20	159284.6	1.029473
96759.37	159880.7	1.029396
96461.35	160357.5	1.029329
96222.93	160893.9	1.029261
95984.51	161430.4	1.029194
95746.09	161907.2	1.029136
95507.66	162443.7	1.029074
95388.45	162980.1	1.029019
95090.43	163397.3	1.028968
94911.61	163874.2	1.028920
94673.19	164291.4	1.028875
94494.37	164708.6	1.028834
94375.16	165185.4	1.028792
94136.73	165602.7	1.028752
93957.91	166019.9	1.028715
93779.09	166377.5	1.028683
93600.27	166735.1	1.028653
93481.06	167092.8	1.028625
93302.24	167390.8	1.028601
93183.03	167748.4	1.028575
93004.21	168046.4	1.028552
92885.00	168344.4	1.028531
92706.18	168582.9	1.028513
92586.97	168880.9	1.028494
92527.36	169238.5	1.028473
EXIT		

PROBLEM 5.7

If we look at the absolute value of the difference between the desired response and the actual response, we get a different performance index. Redo the *RC* filter problem using this new, redefined PI. You will have to change one line in FOFX to form the difference differently:

```
DIFF = CABS(GD(I)-G)
```

Explain your results.

PROBLEM 5.8

The gradient function can often provide significant clues about the gain to use in a gradient optimization solution. Using FOFX (Table 5.7) and RCFILTER (Table 5.6) and the gradient calculation routine (GRAD; see Table 5.2), determine the gradient of FOFX when both resistances are 80 $k\Omega$. Using that computed gradient, determine a value of gain to use in the gradient optimization algorithm. The gain chosen should be such that neither resistance value is changed by more than 5 percent in the first step. Determine the gain value and the resistor that determines the limiting value.

5.10 AN EXTENDED PROBLEM: DESIGNING A MINIMUM-WEIGHT TRANSFORMER

In this section we will spend some time developing the performance index for a more complex problem. If we have an application that needs a special-purpose transformer and if that transformer has to be put aboard an aircraft or a spacecraft, then one very important property of the transformer is how much it weighs, or how much mass it has. A complete study of this sort of problem would encompass more territory than we could cover here. So, if you are an experienced transformer designer, the design here will seem crude. However, the idea of optimizing the design numerically is what we hope to focus on, so let's proceed.

We will assume that we have chosen the particular configuration shown in Figure 5.7. Other configurations might lead to lighter designs, so we should not neglect that possibility. However, to keep things short, we will just try to optimize this configuration, and leave the comparison with other configurations to the interested student.

The transformer consists of two conductor windings wound around a core of some sort of magnetic material. Although there is some insulation, most of the weight is in the magnetic material and the conductors. Our strategy will be to compute the volume of both the magnetic material and the conductors separately and then compute the weight by multiplying by the density of each.

First we compute the volume of the magnetic material, as shown in Figure 5.8. There are four corners, all square, of area X_1^2. Similarly, there are a top and bottom, each with area $X_1 X_2$, and two sides with area $X_1 X_3$ each. Thus, we can get the volume of the magnetic material by adding all these areas to get the area of the "square doughnut," and then multiply by the depth, X_4:

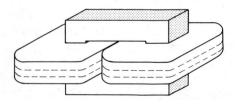

Figure 5.7 A transformer.

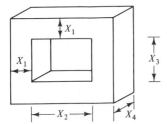

Figure 5.8 The transformer core.

$$\text{Magnetic material volume} = (2X_1 X_2 + 2X_1 X_3 + 4X_1^2)X_4 \qquad (5.10.1)$$

We also need to compute the volume of the wire conductors. Referring to Figure 5.9, we see that we have four corners again, but this time they are rounded. However, we could lump all four corners together, for purposes of computation, and note that they have the area of a circle of radius $X_2/2$. We also need to remember that two windings have to fit into the window area, so only half that area is available for each winding.

This time, we have sides of area $X_2 X_4$, and top and bottom of area $X_1 X_2$. Adding and multiplying by the depth, we get the total volume of the conductors:

$$\text{Conductor volume} = \left(X_2 X_4 + X_1 X_2 + \frac{\pi X_2^2}{4}\right)X_3 \times 2 \qquad \text{(two windings)} \quad (5.10.2)$$

There are a lot of other items of information we need to incorporate into our formulation of the weight. First, we need to multiply magnetic material and conductor volumes by the appropriate densities. We will assume that the magnetic material is some sort of magnetic steel, with a density of 7.8 g/cm³ (or 7800 kg/m³). We also assume that the conductors are copper, with a density of 8.9 g/cm³.

In both the steel, and the copper, it is not true that the material occupies all of the volume we have calculated. The magnetic material is usually laminated, and may take up only about 90 percent of the total volume. We will assume that. We will also assume that the windings take up only about $0.5\pi/4$ (or a little less than 40 percent) of the available volume. Taking all this into account, we have the total mass as the sum of the mass of the magnetic material and of the conductors:

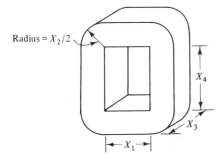

Radius = $X_2/2$

Figure 5.9 The transformer windings.

$$\text{Total mass (kg)} = 8900\left(\frac{0.5\pi}{4}\right) \times \text{Conductor volume}$$

$$+ 7800(0.9) \times \text{Magnetic material volume}$$

$$= 8900\left(\frac{0.5\pi}{4}\right)\left(X_2X_4 + X_1X_2 + \frac{\pi X_2^2}{4}\right)X_3 \times 2$$

$$+ 7800(0.9)(2X_1X_2 + 2X_1X_3 + 4X_1^2)X_4 \qquad (5.10.3)$$

Looking at the expressions for the mass of the magnetic material and of the conductors, we would seem to be able to get a minimum by choosing X_1, X_2, X_3, and X_4 all to be zero. There's got to be something wrong here, and the problem is that we haven't taken everything into account yet. What we have forgotten is to account for the properties of the materials involved. As in any realistic problem, we must take into account the physical limitations inherent in a situation.

For our transformer, we must take into account the physical laws that the device obeys. For example, we can write the voltage across a winding in terms of the rate of change of flux:

$$V(t) = \frac{N d\phi}{dt} \qquad (5.10.4)$$

Further, the flux ϕ is related to the flux density in the magnetic material. If we assume that the flux density is the same anywhere within the magnetic material, then we can say that the flux is the product of flux density and cross-sectional area through which the flux goes. In our case, one leg of the transformer has an area of X_1X_4. Thus we have the relation

$$\phi = BX_1X_4 \qquad (5.10.5)$$

where B = flux density in webers per square meter (Wb/m^2) and X_1X_4 is the cross-sectional area through which the flux flows.

We will assume that the voltage applied to the primary winding is sinusoidal, with

$$V(t) = V_{\text{peak}} \sin (2\pi Ft)$$

With that voltage, we can compute the flux density as a function of time so that we can get at the maximum flux density. However, we need to specify the frequency. In aircraft applications where weight is costly, smaller transformers can be obtained by using higher frequencies. Here we will assume a frequency of 400 Hz. In one of the miniprojects we will examine the same design but at a higher frequency.

$$V(t) = V_{\text{peak}} \sin (2\pi Ft) = \frac{N d\phi}{dt} = NX_1X_4 \frac{dB}{dt} \qquad (5.10.6)$$

Hence

$$B(t) = -\left(\frac{V_{\text{peak}}}{NX_1X_4(2\pi F)}\right) \cos (2\pi Ft)$$

$$= -B_{\text{peak}} \cos (2\pi Ft) \qquad (5.10.7)$$

Thus we have

$$B_{peak} = \frac{V_{peak}}{NX_1X_4(2\pi F)} \qquad (5.10.8)$$

For any magnetic material there is a maximum flux density that the material will support. If that flux density is B_{max}, it would be wise to allow some margin of safety. We will do that by assuming that the largest flux density that we will permit is $0.9B_{max}$. If we operate the transformer so that our peak flux density, B_{peak}, is $0.9B_{max}$, we have

$$0.9B_{max} = \frac{V_{peak}}{NX_1X_4(2\pi F)} \qquad (5.10.9)$$

The equation just developed gives us a relationship between some well-defined quantities like B_{max} (which is a physical property of the material), V_{peak} (which must be a given), and the frequency of operation F. However, N is un-determined.

If we examine the limitations on current in the wire in the windings, we can get N expressed in terms of other (defined) variables. Let us assume that we need to carry a current of 10 A, and that we therefore select no. 10 wire. (Be careful, however; you ought to realize that no. 40 wire, for example, is much smaller, and cannot safely carry 40 A.) The no. 10 wire chosen has a diameter of 0.102 in., or 0.259 cm. The area of each wire is then given as

$$\text{Wire area} = (0.259)^2(0.0001)(\pi/4) \qquad (5.10.10)$$

However, this is just the area of the copper. Earlier, we indicated that we had to make some allowance for the fact that the entire winding space is not taken up by the conductors, and that there is some insulation (and general slack in the windings). The actual wire area per winding will have to be multiplied by the reciprocal of our "winding factor," $0.5\pi/4$. Thus we have

$$\text{Wire area per winding} = (0.259)^2(0.0001)\left(\frac{\pi/4}{0.5\pi/4}\right)$$
$$= (0.259)^2(0.0001)2 \qquad (5.10.11)$$

(Actually, we picked this to work out simply. However, the winding factor is usually between 30 and 40 percent, so our finagling was in the right ballpark.)

Now, we can compute the number of windings that will fit inside the area available, $X_2X_3/2$. The number of turns is

$$N = \frac{X_2X_3/2}{(0.259)^2(0.0001)2}$$
$$= 37,268X_2X_3 \qquad (5.10.12)$$

Now if we use this value in our expression for B_{peak}, we get

$$X_1X_2X_3X_4 = 11.07 \times 10^{-6} \qquad (5.10.13)$$

Here, we have assumed $B_{max} = 1.5$ Wb/m^2 and $V_{peak} = 1000 \cdot \sqrt{2}$.

Summing up, we have the mass of the transformer given by

Mass = Mass of magnetic material + mass of conductors

$$= 28{,}080X_1X_1X_4 + 14{,}040X_1X_2X_4 + 14{,}040X_1X_3X_4 + 3495X_2X_2X_3$$
$$+ 6990X_2X_3X_4 + 6990X_1X_2X_3 \qquad (5.10.14)$$

But the variables must satisfy

$$X_1X_2X_3X_4 = 11.07 \times 10^{-6} \qquad (5.10.15)$$

We could use this relationship to eliminate any one of the four variables: X_1, X_2, X_3, or X_4. We will choose to eliminate X_4, using

$$X_4 = \frac{11.07 \times 10^{-6}}{X_1X_2X_3} \qquad (5.10.16)$$

Substituting this expression for X_4 into the mass, we have

$$\text{Mass} = \frac{.3108X_1}{(X_2X_3)} + \frac{.1554}{X_3} + \frac{.1554}{X_2} + \frac{.0738}{X_1}$$
$$+ 3495X_2X_2X_3 + 6990X_1X_2X_3 \qquad (5.10.17)$$

We can make the substitution for X_4 explicitly as above, or in a program we can calculate X_4 from the other three variables and just work with X_1, X_2, and X_3. In the FOFX subroutine given in Table 5.9 we take the second approach. In the miniprojects at the end of the chapter we will work with this function and get the minimum-mass transformer.

5.11 GENERALIZATIONS AND CONCLUSIONS

By this point we have become familiar with a basic method for solving optimization problems numerically. Further, we have applied that method to a few

TABLE 5.9

```
      SUBROUTINE FOFX(X,N,F)
C
C THIS SUBROUTINE CALCULATES THE WEIGHT OF A TRANSFORMER
C
      DIMENSION X(10)
C
C THE CONSTRAINT IS NOT EXPLICITLY INCLUDED
C
      X(4) = 11.07E-6/(X(1)*X(2)*X(3))
C
      WTMAG = 28080.*X(1)*X(1)*X(4)
      WTMAG = WTMAG + 14040.*X(1)*X(2)*X(4)
      WTMAG = WTMAG + 14040.*X(1)*X(3)*X(4)
      WTCOND = 3495.*X(2)*X(2)*X(3)
      WTCOND = WTCOND + 6990.*X(2)*X(3)*X(4)
      WTCOND = WTCOND + 6990.*X(1)*X(2)*X(3)
      F = WTMAG + WTCOND
      RETURN
      END
```

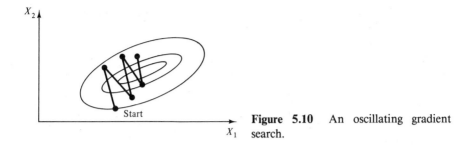

Figure 5.10 An oscillating gradient search.

more or less realistic problems. However, the gradient algorithm is not without its faults. It can be ponderously slow if the performance index has a "skinny" ridge or its higher-dimensional equivalent.

The possibility of slow convergence has led to numerous inquiries into improved and alternate methods of numerical optimization. In this section we will examine some gradient-based techniques that exhibit faster convergence, and we will look at some search techniques that are efficient and popular.

If we think about how a gradient search proceeds, we will usually find behavior such as that shown in Figure 5.10, a two-dimensional example. As a first try at improving the gradient, we can consider a strategy in which we do not always proceed in the direction opposite to the gradient. If we imagine starting at a point, X_0, we can proceed along the gradient-indicated direction until we arrive at a point at which the gradient is perpendicular to the direction traveled. At that point, if we plot the performance index as a function of the distance along the direction traveled, we will find a local minimum at the point of perpendicularity; see Figure 5.11.

Here, we are implying that there is some method available that will permit us to locate that local minimum. Before we continue, try to invent one or two algorithms for determining that local minimum.

One possible method would be to take a small step in the direction toward the minimum. If that step (opposite to the gradient) works, then we could increase the step size and take another step in the same direction. Continuing in this manner, we could continue to increase the step size as long as we were successful. When we take a step that leads to a point at which the PI is larger than the PI at the last point, then we will know that the minimum has been passed, and we will have "bracketed" the solution. At the end of this one-dimensional search,

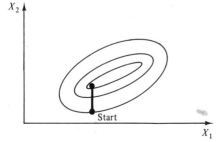

Figure 5.11 A one-dimensional search along the gradient.

we are left with an interval in which we know the local minimum lies, but we still have to locate the local minimum.

The kind of search we are describing above is referred to as a "univariate search," since we are searching in one dimension. There are many techniques for univariate searches, and in Chapter 6 we will look at the one which seems to be the most popular, and (not incidentally) the most efficient.

BIBLIOGRAPHY

G. S. G. Beveridge and R. S. Schechter, *Optimization: Theory and Practice,* McGraw-Hill, New York, 1970.

B. D. Campbell, Grasping the concepts behind optimization methods for control, *Control Engineering,* November 1979, pp. 59–62. (A good, short introduction to the general topic of optimization.)

R. W. Daniels, *An Introduction to Numerical Methods and Optimization Techniques,* North-Holland, New York, 1978 (See Chapter 9 for a good introduction to the gradient method.)

J. E. Dennis, Jr., and R. B. Schnabel, *Numerical Methods for Unconstrained Optimization and Nonlinear Equations,* Prentice-Hall, Englewood Cliffs, NJ, 1983.

R. W. Hamming, *Introduction to Applied Numerical Analysis,* McGraw-Hill, New York, 1977. (A very readable introduction to the concepts and problems in numerical optimization.)

R. L. Zahradnik, *Theory and Techniques of Optimization for Practicing Engineers,* Barnes and Noble, New York, 1971.

MINIPROJECTS

1. Find the dimensions of the minimum-weight transformer. Use the gradient method. You will need to determine a starting value of the parameter vector X (that is, you must find starting values of X_1, X_2 and X_3). Since we are using units of meters, we probably ought to realize that a transformer with a 1-kVA rating would be too large if it were 1 m on a side. Conversely, don't expect dimensions to be less than a centimeter. With this information in mind, do the following.
   (a) Determine a gain value that produces convergence.
   (b) Determine the dimensions of the optimum transformer.
   (c) Try different starting points. You may find that you inadvertently wander into regions with negative parameter values. Determine what happens then, and explain your result.

2. Do the same problem as in Miniproject 1, but rederive the performance index using a frequency of 1 kHz. Eliminate part C of the problem.

3. Using the original performance index for the RC network in Section 5.9, determine gain values that give oscillatory and nonoscillatory approaches to the optimum location. Plot the results in two dimensions. Use different starting values of the parameters and interpret the results.

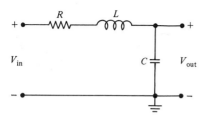

4. Redo the frequency response optimization, but use an *RLC* network as shown in the accompanying figure. Choose a resistance of 1000 Ω, and allow *L* and *C* to vary. Use the performance index developed in the text.

5. Redo the frequency response optimization, but use a general second-order transfer function. Determine the natural frequency and damping ratio that give the best frequency response fit using the performance index developed in the text.

6. Roundoff can influence the accuracy of a computed gradient. In this miniproject do a straightforward gradient optimization of the function

$$F(X) = X_1^2 + 2X_2^2 + \text{constant}$$

Use increments of 0.001 in computing the gradient. Start from (1, 1). Determine the value of the constant that produces total immobility from that starting point with 0.001 increments for gradient computation. Then explain why the immobility occurs, and determine how the region of immobility depends upon the constant for your particular computer system.

Chapter 6

Optimization II

6.1 INTRODUCTION

The gradient method presented in the first optimization chapter is a widely used algorithm not only for the types of optimization we have been discussing, but in many other contexts as well. However, it is just one of many different sorts of optimization algorithms. One other popular type is the class of search algorithms, some of which will be examined in this section.

Let us imagine that we have a function of a single variable to minimize. We will *assume* that the function has only a single minimum, as shown in Figure 6.1. Such a function is called *unimodal*. There are numerous practical situations in which it is reasonable to assume unimodality.

Now, imagine that we have made three measurements of the function. We could have either of the situations depicted in Figure 6.2. The minimum could be in either the left half of the interval or the right half. From these limited measurements we cannot rule out either situation.

The situation changes when we make four measurements of the function. In this situation the two middle data points can be used to make a decision on where the minimum *cannot* be. For example, if $f(X_1) < f(X_2)$, then we can conclude that the minimum cannot possibly lie in the rightmost interval if the function is unimodal. Conversely, if $f(X_2) > f(X_1)$, then the minimum cannot be in the leftmost interval. If we can eliminate an interval that cannot contain the location of the optimum, then by taking more data points we can gradually home in on the location of the optimum, as shown in Figure 6.3.

Narrowing down the range of the location of the optimum should be done as efficiently as possible. If we eliminate an interval, then the remaining interval has one internal data point already taken, as well as two endpoints measured.

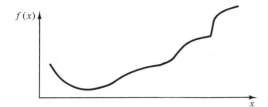

Figure 6.1 A unimodal function.

If we take one more internal data point, we can then eliminate another interval. This process is shown in Figure 6.4.

Let us try to visualize the last few steps of a process that culminates in the situation shown in Figure 6.4, where all of the final subintervals are of equal size. Here we make a decision to eliminate either the leftmost interval or the rightmost interval, depending upon the values of the function at the interior points. We assume that all the subintervals are the same size. Next, we ask ourselves what kind of situation preceded the one we now have. To get to the last, equal subinterval, situation the penultimate situation must have been one of the two shown in Figure 6.5.

The penultimate situation should have been symmetrical. If the length of the equal final subintervals is L, then the last decision eliminates an interval of length L out of a total interval of length $3L$. The step just before eliminates an interval of length $2L$ from a total of $5L$, and the length of the interval eliminated does not depend upon which side the interval happened to be.

If we continue this process, we will find that the way we arrive at an interval of length $5L$ is to eliminate an interval of $3L$ out of one of $8L$ (see Figure 6.6). Following this process along, we find that the length of the interval in the kth last step (call it L_k) is given by

$$L_k = L_{k-1} + L_{k-2} \tag{6.1.1}$$

This particular relation holds for the last three intervals of 2, 1, and 1. Then, the intervals can be calculated as

$$
\begin{array}{ll}
2 = 1 + 1 & \text{1-1} \\
3 = 2 + 1 & \text{1-1-1} \\
5 = 3 + 2 & \text{2-1-2} \\
8 = 5 + 3 & \text{3-2-3} \\
\text{and so forth}
\end{array}
$$

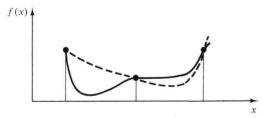

Figure 6.2 Two unimodal functions that yield the same values at three points.

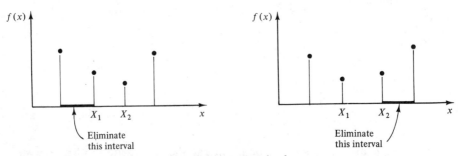

Figure 6.3 The decisions made based on measured values.

This particular sequence of numbers, known as the Fibonacci sequence, has become very well known. We can construct an entire search process from the numbers in the Fibonacci sequence, but it may not be worth the special effort to calculate all the terms in the sequence. In the next section we will examine a limiting case of the Fibonacci search, a method that is frequently used [1, 2].

6.2 THE GOLDEN SECTION SEARCH

In the limit, successive Fibonacci numbers tend toward a particular ratio. To see this, examine several terms in the sequence, as follows:

$$2/1 = 2.0$$
$$3/2 = 1.5$$
$$5/3 = 1.6667$$
$$8/5 = 1.6000$$
$$13/8 = 1.6250$$
$$21/13 = 1.6154$$
$$34/21 = 1.6190$$

The limiting ratio is usually referred to as the Golden Ratio. It was originally a ratio deemed especially pleasing, esthetically, by classical Greeks. This ratio is the ratio of width to height, for example, in many classical Greek buildings. The exact value is one-half of the sum of unity and the square root of 5; that is,

$$\text{Golden Ratio} = \frac{1 + \sqrt{5}}{2} \qquad (6.2.1)$$

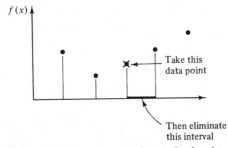

Figure 6.4 How the next data point is taken.

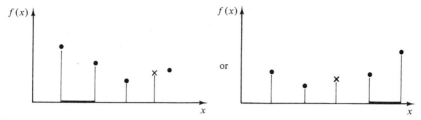

Figure 6.5 Two ways in which the previous situation could have occurred.

This value comes out of a solution to the difference equation for the lengths. Assume that $L_k = r^k$ and we have

$$L_k = L_{k-1} + L_{k-2} \tag{6.2.2}$$
$$r^k = r^{k-1} + r^{k-2} \tag{6.2.3}$$

or

$$r^k - r^{k-1} - r^{k-2} = 0$$
$$r^2 - r - 1 = 0$$
$$r = \frac{1 \pm \sqrt{5}}{2} \tag{6.2.4}$$

The actual solution for L_k is a linear combination of 1.618 and 0.618 raised to the kth power. In the limit, $L(k) = $ Constant $\times 1.618^k$, and the growing term will ultimately dominate in L_k. Eventually, successive interval sizes will be in the ratio of 1.618034:1. (In the few computations earlier, that ratio was approached very quickly, within five or ten iterations.)

Instead of bothering to compute all of the Fibonacci intervals, one alternative approach is to use the "Golden Section" division for the interval for every iteration. In this modification of the algorithm every step looks like the one shown in Figure 6.7. At each step, the total interval is reduced by a factor of 0.618034. At the kth step, the interval is reduced by the kth power of 0.618034.

PROBLEM 6.1

Determine how many steps of the Golden Section search are needed to bring the area of uncertainty in the location of the optimum to within 1 percent of the original interval. Do again for 0.1 percent.

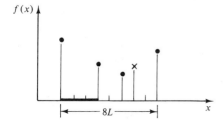

Figure 6.6 Data points in an interval of $8L$.

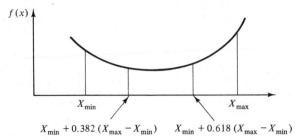

Figure 6.7 How intervals are chosen in the Golden Section search.

PROBLEM 6.2

In doing root locus analysis in control systems problems, finding the break-away point (the point at which the locus leaves the real axis) is a numerical problem that needs to be addressed. In this system, it is known that there

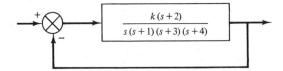

is a breakaway point between 0 and -1. Further, it is known that the breakaway point occurs where the function, $F(s)$, has a maximum:

$$F(s) = -\frac{s(s+1)(s+3)(s+4)}{(s+2)} \qquad -1 < s < 0$$

Using a Golden Section search, determine the location of the breakaway point between 0 and -1. Your value should be correct to three significant figures.

ALGORITHM

GOLDEN SECTION SEARCH

The Golden Section search algorithm can be used to find the location of the minimum (maximum) of a unimodal function of a single variable. Steps in the algorithm are as follows:

1. Determine an initial search interval. Call the left end of the search interval X_{min}, and the right end of the search interval X_{max}.
2. Calculate two points in the interior of the search interval, X_{left} and X_{right}. These two points are located at

$$X_{left} = X_{min} + 0.381966(X_{max} - X_{min})$$
$$X_{right} = X_{min} + 0.618034(X_{max} - X_{min})$$

3. Evaluate the function $F(X)$, and let

$$F_{left} = F(X_{left})$$
$$F_{right} = F(X_{right})$$

4. Redefine either X_{min} or X_{max}.

If $F_{left} > F_{right}$, choose a new $X_{min} = X_{left}$.

If $F_{right} > F_{left}$, choose a new $X_{max} = X_{right}$.

5. Go back to step 2, and repeat steps 2 through 5 until the interval from X_{min} to X_{max} has been shrunk to an acceptable size. Take the value producing a minimum function value to be the midpoint of the final interval.

PROBLEM 6.3

A student has measured the step response of a system. The data taken is given below ($\Delta t = 0.2$ s).

Time	Response
0.0	0
0.2	0.2
0.4	0.35
0.6	0.47
0.8	0.59
1.0	0.67
1.2	0.72
1.4	0.79
1.6	0.82
1.8	0.85
2.0	0.90

The student suspects that the data fits the form

$$1 - e^{-k\,\Delta t/\tau}$$

where $\tau =$ Time constant and $k = 0, 1, \ldots, 10$. The student wants to determine the time constant that gives the best "fit" to the data, and has devised this function as a measure of how well the data fits:

$$f(\tau) = \sum_{k=0}^{10} |\text{Response } (k\,\Delta t) - (1 - e^{-k\,\Delta t/\tau})|^N$$

When this function is a minimum, the data fits best.

(a) Using values of N of 1 and 2, determine the "best" time constant (using each definition of "best"). Use a Golden Section search. Find the time constant to within 0.01 s.

(b) For the best fit, determine the value of the function.

6.3 DIRECTIONAL MINIMIZATION AND THE FLETCHER-POWELL METHOD

The search methods presented in the first two sections of this chapter are efficient methods for univariate searches. They are not, however, easily extended or mod-

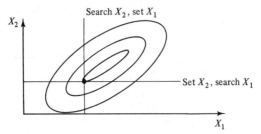

Figure 6.8 Searching X_1 and X_2 separately.

ified to handle functions of several variables. One might be tempted to search successive variables separately, but inevitably the search proceeds laboriously along skinny ridges. Figure 6.8 shows such a situation. After searching the first variable and landing on the ridge, searching the second variable will yield little improvement.

In this section we will digress and consider a "mixed-mode" optimization method, one that combines search and gradient methods. After consideration of this method (the Fletcher-Powell method), we will examine "pure" multivariate search methods. Those methods are more heuristic, and often are without any analytical foundation. They are, however, very effective and widely used. We will consider two search methods after examining the Fletcher-Powell method, first examining behavior for a quadratic multivariable function.

If we are trying to find the minimum of a function that can be represented by a quadratic function, we could use analytical techniques to find the minimum. If the function is

$$f(\mathbf{X}) = (\mathbf{X}^T A \mathbf{X}) + (\mathbf{B}^T \mathbf{X}) + C \tag{6.3.1}$$

Then, we have previously determined (in Chapter 5, Section 5.7) that the location of the minimum is at

$$\mathbf{X}_{opt} = -\tfrac{1}{2}A^{-1}\mathbf{B} \tag{6.3.2}$$

We can compare the direction we would like to move and the direction of the gradient. Say we are at an initial location, $\mathbf{X}_0$. From the quadratic expression we can compute the gradient and form the difference between the minimizing point, $\mathbf{X}_m$, and the initial point, $\mathbf{X}_0$.

$$\mathbf{X}_{opt} = -\tfrac{1}{2}A^{-1}\mathbf{B} \tag{6.3.3}$$
$$\mathbf{X}_{opt} - \mathbf{X}_0 = -\tfrac{1}{2}A^{-1}\mathbf{B} - A^{-1}A\mathbf{X}_0$$
$$= -\tfrac{1}{2}A^{-1}[\mathbf{B} + A\mathbf{X}_0]$$
$$= -\tfrac{1}{2}A^{-1}\nabla f(\mathbf{X}_0) \tag{6.3.4}$$

The direction to move, $\mathbf{X}_{opt} \rightarrow \mathbf{X}_0$, depends upon the gradient direction, but is not in the same direction. The inverse of A acts to rotate the direction in the parameter space.

Essentially, the surface is modeled with a quadratic function that includes A (a square matrix), $\mathbf{B}$ (a vector), and C (a scalar) as parameters. If we know A

and **B,** we could then compute $\mathbf{X}_{\mathrm{opt}}$. If we do not know A and **B,** it might still be possible to make measurements that permit us to compute A and **B.**

Consider how A and **B** could be measured. The gradient, $\nabla(f(\mathbf{X}))$, is given by

$$\nabla(f(\mathbf{X})) = 2A\mathbf{X} + \mathbf{B} \qquad (6.3.5)$$

The expression for the gradient depends linearly upon the parameter vector, **X.** If **X** is varied, we can compute the matrix, A. If one component of **X** is varied, we can compute a column of A. Imagine that one component of **X** is increased an amount $\mathbf{DX}_i$ (the ith component). Then we would have

$$\nabla(f(\mathbf{X} + \mathbf{DX}_i)) - \nabla(f(\mathbf{X})) = 2A\,\mathbf{DX}_i \qquad (6.3.6)$$

If $\mathbf{DX}_i$ is a vector with only one nonzero entry in the ith location, then this operation will "pick out" the ith column in A and the resulting difference in the gradients at $\mathbf{X} + \mathbf{DX}_i$ and **X** will be proportional to the ith column in A.

We can also compute the vector **B.** If the parameter vector **X** is set to zero, then

$$\nabla(f(\mathbf{0})) = \mathbf{B} \qquad (6.3.7)$$

However, if we are attempting to locate a local minimum of $f(\mathbf{X})$, the quadratic representation may only be valid locally near the present value of **X.** Moving to $\mathbf{X} = \mathbf{0}$ and evaluating the gradient may take **X** beyond the area in which the quadratic representation is valid, and measuring $\nabla(f(\mathbf{0}))$ may not be a realistic way of measuring **B.** A simple way around the difficulty is to redefine the coordinate system in the parameter (**X**) space so that the origin is at the present point, **X.** Then, in that coordinate system the vector **B** would just be the gradient at **X.**

However we define the coordinate system for the problem, there is a computational problem with the approach just outlined. Computation of the gradient is really computation of a set of derivatives, and we probably would do that by computing the function at different points and taking differences. Taking differences is a process that is numerically "noisy." If we take a difference between two closely spaced function values, for example, the difference will be small and roundoff will be accentuated. Thus a numerical computation of a gradient will necessarily have inaccuracies. Those inaccuracies will be accentuated if we then take the difference between two computed gradients as we would have to do to compute A in the quadratic representation of the function $f(\mathbf{X})$.

Another possible numerical problem is that the quadratic representation is probably only an approximation to the actual function. There is no a priori guarantee that the quadratic model will be a good one. Finally, even if A were known accurately, it is still necessary to invert A, and matrix inversion will inevitably introduce still more inaccuracy in the computed inverse of A. Thus, although this approach is tempting, there is a strong possibility that it might not work well because of numerical problems.

ALGORITHM

QUADRATIC JUMP

If we have a value of a parameter vector that is known to be relatively near to the optimum value, then it is reasonable to assume that the performance index is well approximated by a quadratic expression near the optimum, and to try to take advantage of that behavior in order to find the optimum location. The quadratic jump method might be used in that situation.

The method assumes a quadratic representation as follows:

$$f(\mathbf{X}) = \mathbf{X}^{\mathrm{T}} A \mathbf{X} + \mathbf{b}^{\mathrm{T}} \mathbf{X} + C$$

The gradient of the quadratic function is

$$\nabla f(\mathbf{X}) = A \mathbf{X} + \mathbf{b}$$

The steps in the algorithm are as follows:

1. Evaluate the gradient at the starting point, X_0.
2. Determine the Jacobian matrix by incrementing all components of the parameter vector individually:

$$A = \{J[f(\mathbf{X})]_{ij}\} = \left\{ \frac{\partial^2 f}{\partial X_i\, \partial X_j} \right\} \simeq \frac{\nabla f(\mathbf{X} + \mathbf{D}\mathbf{X}_j)|_i - \nabla f(\mathbf{X})|_i}{\mathbf{D}\mathbf{X}_j}$$

3. Move to the new location in the parameter space:

$$\mathbf{X}_{\mathrm{opt}} = \mathbf{X}_0 - \tfrac{1}{2} A^{-1} \nabla f(\mathbf{X}_0)$$

The new location found by this method may not necessarily be the optimum if a quadratic representation is not locally "tight." A fourth possible step is to check the value of the gradient at this point, to see whether it is sufficiently close to the zero vector.

PROBLEM 6.4

Determine the local quadratic representations of the functions below, and do one step of the quadratic jump method. Determine how close the method comes when starting at the indicated starting points.

(a) $f_A(\mathbf{X}) = [\sin(X_1 - X_2)]^2 + (X_1 - 1)^2$
   Start at $X_1 = 0, X_2 = 1$ Min at $X_1 = 1, X_2 = 1$
   and $X_1 = 0.5, X_2 = 0.5$

(b) $f_B(\mathbf{X}) = [\tan(X_1 - X_2)]^2 + (X_2 - 1)^2$
   Start at $X_1 = 0, X_2 = 1$
   $X_1 = 0.5, X_2 = 0.5$

In 1963 Fletcher and Powell published an article describing a method that takes advantage of the local "quadraticity" of most performance indices [3]. Moreover, their iterative method contains a matrix that converges to the inverse

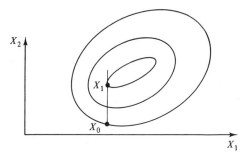

Figure 6.9 Directional optimization.

of A in the quadratic model without ever explicitly taking a matrix inverse. In the next section, we will describe the Fletcher-Powell method and discuss the peculiarities of its various steps. In the remainder of this section we will finish the discussion of the directional optimizer, with an eye toward the Fletcher-Powell method.

We begin by considering how to do a directional optimization. Starting at X_0, we need to locate X_1 as shown in Figure 6.9. By computing the gradient we can find the direction to move. However, the distance to the directional minimum point, X_1, cannot be found without more information. One feasible way to find X_1 is simply to search for a minimum along the direction of the gradient.

In the directional search for the minimum there are no a priori bounds on **X,** so we cannot start out with a Fibonacci search, for example. Without knowing anything about the location of the direction minimum, there is a distinct danger that the search method could be too "timid." One "bold" measure is to accelerate changes in **X** as long as improvement is found, decelerating and reversing direction when the performance index increases. Another strategy is to search in one direction, accelerating the step size until it is clear that the minimum has been passed by. Figure 6.10 shows how such a search could proceed. Once we have located an interval containing the minimum, then a Golden Section search could be conducted along the search direction within the known minimum-containing interval. Table 6.1 gives the directional optimization subroutine.

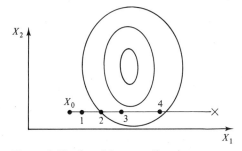

Figure 6.10 Searching one direction.

TABLE 6.1

```
C
C********************************************************
C
      SUBROUTINE DIROPT (X,D,N,A)
      DIMENSION  X(20), D(20), XTEST(20)
C
C     X = PARAMETER VECTOR
C     D = Vector defining search direction (gradient)
C
      ALPHAL = 0.
      ALPHAR = A
C
C*******SAVE X(old) vector and find initial F(X) values*********
C
      CALL FOFX (X,N,FOLD)
C
C**********Find X(new)*************
C
          DO 10 I = 1,N
               XTEST(I) = X(I) + ALPHAR*D(I)
   10     CONTINUE
          CALL FOFX (XTEST,N,FNEW)
      DO 100 IK = 1,20
C
C*****Find new F(X) value and compare with old F(X) value********
C     If F(X)new > F(X)old reverse direction and cut distance scale.
C     If F(X)new < F(X)old double distance multiplier and continue.
C
          IF (FNEW .LT. FOLD) THEN
               ALPHAR = 2.*ALPHAR
               FOLD = FNEW
            DO 20 I = 1,N
               XTEST(I) = X(I) + ALPHAR*D(I)
   20          CONTINUE
            CALL FOFX(XTEST,N,FNEW)
          ELSE
               IF (IK .GT. 2) ALPHAL = ALPHAR/4.
      ALPHA1 = ALPHAL + (.381966)*(ALPHAR-ALPHAL)
      ALPHA2 = ALPHAL + (.618034)*(ALPHAR-ALPHAL)
      DO 60 J=1,N
           XTEST(J) = X(J) + ALPHA1*D(J)
   60 CONTINUE
      CALL FOFX(XTEST,N,F1)
      DO 70 J=1,N
           XTEST(J) = X(J) + ALPHA2*D(J)
   70 CONTINUE
      CALL FOFX(XTEST,N,F2)
      DO 99 I=1,12
        IF (F1 .GT. F2) THEN
               ALPHAL = ALPHA1
               ALPHA1 = ALPHA2
               F1 = F2
               ALPHA2 = ALPHAL + (.618034)*(ALPHAR-ALPHAL)
          DO 80 J=1,N
               XTEST(J) = X(J) + ALPHA2*D(J)
   80     CONTINUE
          CALL FOFX(XTEST,N,F2)
        ELSE
               ALPHAR = ALPHA2
               ALPHA2 = ALPHA1
               F2 = F1
               ALPHA1 = ALPHAL + (.381966)*(ALPHAR-ALPHAL)
```

TABLE 6.1 (continued)

```
      DO 90 J=1,N
            XTEST(J) = X(J) + ALPHA1*D(J)
  90 CONTINUE
      CALL FOFX(XTEST,N,F1)
         ENDIF
  99 CONTINUE
      DO 200 J = 1,N
      X(J) = X(J) +   (ALPHA1+ALPHA2)*D(J)/2.
 200 CONTINUE
         RETURN
         ENDIF
 100 CONTINUE
      RETURN
      END
```

EXAMPLE 6.1

Below we have a function and the gradient of the function:

$$f(\mathbf{X}) = 10(X_1 - 2)^2 + 3(X_2 - 3)^2$$

$$\nabla f(\mathbf{X}) = \begin{bmatrix} 20(X_1 - 2) \\ 6(X_2 - 3) \end{bmatrix}$$

If we want to optimize the function along the direction of the negative gradient, we find the minimum, with respect to α, of the function

$$f(\mathbf{X} - \alpha\nabla f(\mathbf{X})) = 10[(X_1 - \alpha 20(X_1 - 2))^2 + 6(X_2 - \alpha(X_2 - 3))^2]$$

ALGORITHM

DIRECTIONAL SEARCH

One directional search algorithm searches along the negative gradient direction until an interval can be established that contains the minimum. This is followed by a Golden Section search within the minimum containing interval until the minimum is found to acceptable accuracy.

1. Establish a starting parameter vector, $\mathbf{X}_0$.
2. Establish a starting gain or step size to proceed along the negative gradient direction.
3. Evaluate the function at $\mathbf{X}_0$, obtaining $F(\mathbf{X}_0)$.
4. Move in the direction of the negative gradient.

$$\mathbf{X} = \mathbf{X}_0 - \alpha\nabla f(\mathbf{X}_0) \qquad \mathbf{X}_1 \text{ yields a minimum}$$

5. Evaluate $F(\mathbf{X}_1)$. If $F(\mathbf{X}_1)$ is larger than the starting function value, $F(\mathbf{X}_0)$, then the step size is too large. Go back to step 2 and use a smaller step size.
6. Double the value of gain in the gradient step.
7. Evaluate $\nabla(F(\mathbf{X}_1))$.
8. Move in the direction of the negative gradient.

$$\mathbf{X} = \mathbf{X}_1 - \alpha\nabla f(\mathbf{X}_1) \qquad \mathbf{X}_2 \text{ yields a minimum}$$

9. Evaluate $F(\mathbf{X}_2)$. If $F(\mathbf{X}_2) > F(\mathbf{X}_1)$, then the minimum lies between $\mathbf{X}_0$ and $\mathbf{X}_2$. Otherwise, continue the search, going back to step 3, using $\mathbf{X}_1$ as $\mathbf{X}_0$, and $\mathbf{X}_2$ as $\mathbf{X}_1$:

$$\mathbf{X}_0 \leftarrow \mathbf{X}_1$$
$$\mathbf{X}_1 \leftarrow \mathbf{X}_2$$

10. Do a Golden Section search between $\mathbf{X}_0$ and $\mathbf{X}_2$. In other words, search between 0 and 1 in α.

PROBLEM 6.5

Minimize the two-dimensional function:

$$f(\mathbf{X}) = (X_1 - 5)^2 + 10(X_2 - 7)^2$$

Start at the point (0, 0) and move along the 45° line, where $X_1 = X_2$. Do the minimization analytically and numerically.

EXAMPLE 6.2

A recurring problem is to find the best match between a set of data and the performance of a system. In this example we consider matching the time constant and DC gain of a model system to produce the best match between the model's step response and the step response measurements.

The data for the step response is as follows:

Time	Response
0.0	0.0
2.0	3.4
4.0	5.9
6.0	8.4
8.0	10.3
10.0	12.2
12.0	13.7
14.0	14.7
16.0	15.9
18.0	16.8
20.0	17.5

Clearly, in 20 seconds this system has not reached steady state, so we have no good idea of what the DC gain really is (although it looks to be somewhere in the vicinity of 20). The problem we have is to determine both the DC gain and the time constant.

To solve the problem, we devise a function that measures how far off the DC gain, X_2, and the time constant, X_1, are by comparing a computed response to the measured response. (Compare this example to Problem 6.3.) The expression used is incorporated into an FOFX subroutine as follows:

```
C
      SUBROUTINE FOFX (X,N,F)
      DIMENSION X(10)
      COMMON/DATA/TIME(100),TEMP(100),NTEMP
```

```
C
      F = 0.
      DO 100 I = 1,NTEMP
         AT = X(1)*TIME(I)
         RESPONSE = X(2)*(1.-EXP(-AT))
         F = F + (RESPONSE-TEMP(I))**2
  100 CONTINUE
      RETURN
      END
C
```

In this formulation the data would have to be stored in the COMMONed arrays, TIME and TEMP (for temperature), and passed to FOFX through the labeled COMMON block, DATA.

Running a pattern search with this function produces values of 0.080377 for X_1 and 21.93 for X_2. (A value of 0.08 for X_1 corresponds to a time constant of 12.5 s the way we formulated FOFX.) The minimum FOFX is 0.0876, which would indicate fairly good fit for the values found for X_1 and X_2.

PROBLEM 6.6

Modify FOFX in Example 6.2 to compute the performance index as an accumulation of the absolute value of differences.

This theme of devising model systems that mimic measured performance is continued in Miniproject 5 at the end of this chapter. In the meantime, we will move on to the Fletcher-Powell method.

6.4 THE FLETCHER-POWELL METHOD

The Fletcher-Powell method begins with a directional minimization. First, at some starting point, the gradient of the performance index is evaluated. Then a directional minimization is performed moving along the gradient's direction. The later steps in the algorithm are designed to produce matrices that converge to the appropriate matrices in the quadratic model. A flowchart of the method is given in Figure 6.11.

ALGORITHM

FLETCHER-POWELL OPTIMIZATION METHOD

The steps in the Fletcher-Powell method are as follows:

1. Choose a starting parameter vector $\mathbf{X}_0$ and let

$$G_0 = I \quad \text{(identity matrix)}$$

2. Search in the direction, $\mathbf{d}_0$, for a minimum.

$$\mathbf{d}_0 = -G_0 \nabla^0 \qquad \nabla^0 = \nabla f(\mathbf{X}_0)$$
$$\mathbf{X}_1 = \mathbf{X}_0 + \alpha_0 \mathbf{d}_0 \qquad \alpha_0 \text{ minimizes } f(\mathbf{X}_0 + \alpha_0 \mathbf{d}_0)$$

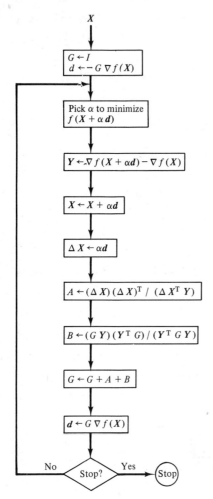

Figure 6.11 Flowchart for the Fletcher-Powell algorithm.

3. Let

$$\mathbf{Y}_0 = \nabla^1 - \nabla^0 \qquad \Delta\mathbf{X}_0 = \mathbf{X}_1 - \mathbf{X}_0 = \alpha_0\mathbf{d}_0$$

4. Update the G matrix:

$$G_1 = G_0 + A_0 + B_0$$
$$A_0 = \frac{(\Delta\mathbf{X}_0)(\Delta\mathbf{X}_0)^{\mathrm{T}}}{(\Delta\mathbf{X}_0)^{\mathrm{T}}\mathbf{Y}_0}$$
$$B_0 = \frac{-(G_0\mathbf{Y}_0)\mathbf{Y}_0^{\mathrm{T}}G_0}{(\mathbf{Y}_0^{\mathrm{T}}G_0\mathbf{Y}_0)}$$

5. Compute a new vector for directional search (not necessarily in the direction opposite to the gradient:

$$\mathbf{d}_1 = -G_1\nabla^1$$

6. Sequence through the steps below until some stopping criterion is satisfied:

$$\mathbf{d}_k = -G_k \nabla^k$$
$$\Delta \mathbf{X}_k = \alpha_k \mathbf{d}_k, \qquad \nabla^{k+1} = \nabla f(\mathbf{X}_{k+1})$$
$$\mathbf{Y}_k = \nabla^{k+1} - \nabla^k$$
$$A_k = \frac{\Delta \mathbf{X}_k \Delta \mathbf{X}_k^T}{\Delta \mathbf{X}_k^T \mathbf{Y}_k}$$
$$B_k = \frac{G_k \mathbf{Y}_k \mathbf{Y}_k^T G_k}{\mathbf{Y}_k^T G_k \mathbf{Y}_k}$$
$$G_{k+1} = G_k + A_k + B_k$$

If we imagine a two-dimensional problem, just doing a directional optimization in the direction of the gradient would produce results of the sort shown in Figure 6.12. Once the directional minimum along the gradient is found, the next gradient would be perpendicular to the previous one, so successive directions would be at 90°. The Fletcher-Powell method uses information that is gathered in the search to modify this situation so that successive directions move more directly to the location of the minimum.

When we begin an optimization, we probably have only the information about the gradient at the starting point. Rather than trying to gather the necessary information locally around the starting point, the Fletcher-Powell algorithm takes information as it is generated and modifies the direction of movement to the direction calculated above as $\mathbf{X}_m \rightarrow \mathbf{X}_0$. In the process, the matrix G converges to the inverse of A without ever having to compute a matrix inverse explicitly.

We will not present a proof of the convergence of the method here. The original paper by Fletcher and Powell [5] discusses the convergence properties and gives an example of performance of the algorithm on the Rosenbrock function (discussed in Chapter 5). Chapter 4 of Bunday [1] presents a discussion of the method and a comparison with other methods, particularly the Fletcher-Reeve method.

In general, the Fletcher-Powell method works considerably faster than a "pure" gradient method, although it must be classed as a gradient-based method. Its good performance depends upon the performance index being able to be modeled with a quadratic model at the minimum. That is not always possible—for example, in constrained minimization with inequality constraints. Still, the method is popular, and its popularity derives from its generally good performance. In the next section we will begin examining search methods that are not gradient-based, but are more heuristic.

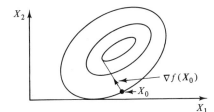

Figure 6.12 Directional optimization along the gradient.

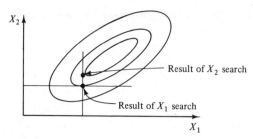

Figure 6.13 Searching X_1 and X_2 separately.

6.5 MULTIDIMENSIONAL SEARCH METHODS: THE SIMPLEX METHOD

The Golden Section and Fibonacci search methods are excellent for single-variable problems. However, they are not appropriate for multidimensional searches, at least without some modification. Consider how we might try to use a one-dimensional search method. The most obvious (and naive) approach would be to set all variables at midpoint values, and then sequentially search for a minimum varying only the first parameter, then the second, and so forth. Figure 6.13 shows the kind of situation that can develop. If we search over X_1, and then set that optimum value for X_1, a search over X_2 will not yield any improvement. Clearly, the technique will fail on this sort of problem, and some other approach or modification will be necessary.

Any search algorithm must clearly take cognizance of the multidimensional aspect of the search. Trying to extend any one dimensional search technique will encounter the same difficulty outlined above. In this section we will discuss one of several search techniques that have proved out well in many practical problems.

Imagine that we have a two-dimensional function to be minimized as depicted in Figure 6.14. Imagine taking three measurements at points on an equilateral triangle in the two-dimensional parameter space (the three points X_1, X_2, and X_3 in Figure 6.14). One possible strategy is to notice that the function at the point X_2 has a larger (and therefore less desirable) value than at the other two points. One intuitively appealing strategy is to "reflect" the point X_2 through the line joining X_1 and X_3, producing a new equilateral triangle as shown in Figure 6.15.

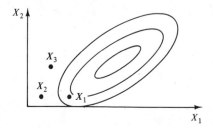

Figure 6.14 Taking three measurements of a function.

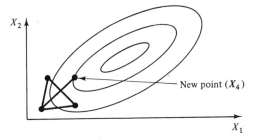

Figure 6.15 Choosing a new point to measure.

After generating the new point, X_4, and noting that we still have an equilateral triangle, we can adopt the same strategy; that is, we examine the value of the function at all current points (throwing away the old point, X_2) and reflect the worst point through the line joining the other two points.

This process can be continued. There might arise some complications, however. For example, a reflection might take us back to a point we had previously thrown out. See Figure 6.16 for an example of this sort of behavior. Figure 6.16 should make it clear that we can have oscillatory behavior if we apply this process blindly. Clearly, we need to allow for some modification in strategy.

The strategy we adopt is as follows:

1. Normally, the worst point (highest function value) is discarded and reflected through the simplex to generate a new point.
2. If oscillations are encountered (detected by noting that we arrive back at a previous point), then the second worst point is discarded and a new point generated by reflection through the simplex.

With this modification the simplex method will not get stranded when the simplex tends to straddle a ridge. (The second rule above is discussed in Rao [4] and in Beveridge and Schechter [5].) However, one problem still remains. In the "end game," as the simplex gets to the optimum location it is possible for more extended cycling to occur. (This cycling may not be simple oscillation back and forth between two points.) The possibility of extended cycling can be avoided if we keep track of the "age" of each point in the simplex. If a point in the simplex gets old and appears continually through a large number of computations, then

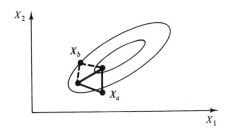

Figure 6.16 Oscillations between X_a and X_b.

our algorithm should suspect that cycling is occurring and should terminate the search.

If the parameter space has more than two dimensions, then essentially the same procedure is followed. In three dimensions the basic figure is a tetrahedron (instead of an equilateral triangle). In higher dimensions, a regular figure frequently referred to as a *polytope* is used (Gill, Murray, and Wright [6]). (The expressions for generating a polytope are given in the separate description of the simplex algorithm.)

Finally, just getting to the point where the simplex oscillates around the optimum location will not necessarily be adequate. If the dimensions of the simplex are too large, then a workable strategy is to shrink the simplex and to search again, shrinking and searching until adequate accuracy has been achieved.

ALGORITHM

SIMPLEX SEARCH MINIMIZATION

The simplex algorithm is used to find the location of a minimum in a (multidimensional) parameter space. The steps in the algorithm are as follows.

1. Determine a starting location, X_0, in the parameter space.
2. Using a "scale factor", a, determine the location of the $N + 1$ points in the polytope. The factor, "a", determines the size of the polytope.

Point	N coordinates of the simplex points relative to X_0							
1	0	0	0	0	•	•	•	0
2	p	q	q	q	•	•	•	q
3	q	p	q	q	•	•	•	q
4	q	q	p	q	•	•	•	q
⋮								
N	q	q	q	q	•	•	p	q
$N + 1$	q	q	q	q	•	•	q	p

where

$$p = \frac{a}{N\sqrt{2}}(N - 1 + \sqrt{N + 1})$$

$$q = \frac{a}{N\sqrt{2}}(-1 + \sqrt{N + 1})$$

3. Find the point in the simplex that has the smallest function value. Call that point X_{min}.
4. Discard X_{min} from the simplex, and add a new point to the simplex by reflecting X_{min}; that is, add the point

$$X_{ref} = \left[\frac{2}{N}\left(\sum_{k=1}^{N+1} X_k - X_{min}\right)\right] - X_{min}$$

5. Continue the search, evaluating the function at all points on the simplex, following the rules:

Rule 1. In general, a new point is generated by finding X_{min} as in step 3 above, removing X_{min} from the simplex and generating a new point by reflection, as in step 4 above. Exceptions to this rule are contained in Rules 2 and 3.

Rule 2. If Rule 1 causes a return to a point just vacated, then instead of discarding X_{min}, determine the point with the next lowest function value, and discard it. Generate a new point to add to the simplex by reflecting the discarded point as in step 4 above.

Rule 3. If application of Rules 1 and 2 cause a return to any point previously vacated prior to the previous step, then either quit (if adequate accuracy has been achieved) or reduce the size of the simplex.

As a point of information, Rule 2 is intended to prevent the kind of oscillation shown in Figure 6.16. Rule 3 prevents extended oscillations involving more than two points in a sequence.

6.6 PATTERN SEARCH: AN ACCELERATED SEARCH METHOD

The method of Hooke and Jeeves otherwise known as the pattern search method, is a method that uses two distinct stages repetitively [1, 2, 5]. Each iteration of the method starts first with a local exploration in the parameter space. Then, after determining a local best point, the method takes a "pattern move." The algorithm is designed so that successive pattern move takes progressively larger steps if those steps seem to go in a direction of improvement (in the sense that the function gets to be less).

Figure 6.17 shows how we would conduct a local exploration in a two-dimensional parameter space. In the example shown, the rightmost point gave

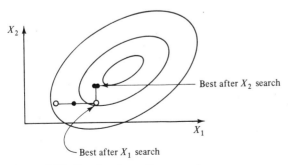

Figure 6.17 Starting the pattern search.

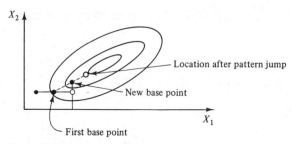

Figure 6.18 A pattern move.

the most improvement when X_1 was varied, and the topmost point gave the most improvement when X_2 was varied. If the situation in Figure 6.17 is the start of a search, the starting point serves as a base point. After the local exploration, we attempt to get an improvement in the function by moving away from the base point in the direction of most improvement. We figure the difference between the most improved point and the base point, and move that distance again as shown.

If the result of our leap is successful (the function is less at that point) then we do another local exploration, getting a new "most improved point." Then we do another jump, again taking the vector from the last base point to the most improved point in the latest local exploration. At each step the most improved point becomes a new base point to use in making the pattern jump as in Figure 6.18.

If we continue this process, we find that the pattern jump gets larger as long as improvement is found. Several steps in the process show how that comes about. The net result is that the pattern search algorithm accelerates movement whenever things continue to go well in a particular direction. (And those directions are not just at particular angles.) Eventually, however, the process can take a jump that does not improve the function. When that happens, we retreat to the last base point, and begin local exploration again. That effectively brings the length of the pattern jump back to its original value, from which point it can begin to accelerate again. Figure 6.19 shows how that point can be reached. The next step will take us well beyond the optimum.

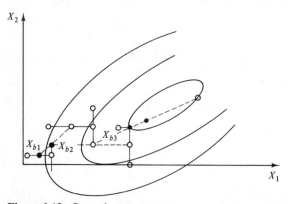

Figure 6.19 Several pattern moves.

Eventually, the search process must reach a point at which no further improvement is obtained in the local exploration. When that happens, the pattern search can produce no further improvement because the local exploratory pattern is probably straddling the optimum. To get further improvement we can diminish the size of the search steps in the local exploration. Usually reduction by a factor of ten works well, and the pattern search can be started again to refine the estimate of the optimum location. Then step size can be decreased and the process repeated as long as the user wants.

ALGORITHM

PATTERN SEARCH: METHOD OF HOOKE AND JEEVES

The pattern search algorithm can be used to find the minimum of a function of N dimensions. Steps in the algorithm are as follows:

1. Determine a starting location, X_0, in the parameter space.
2. Determine increments to be used in searching each of the N parameters. Call these DX_i ($i = 1, \ldots, N$).
3. Define a "base point", X_{base}, as X_0 (to start).
4. Do a local exploration of the parameter space. Increment each variable by 0, $+DX_i$, and $-DX_i$. For each component of the parameter vector, retain the component value producing the smallest (best) function value. Call this new point X_1.
5. If X_1 is the "center" point (all zero increments), then no improvement in the function has been found, and either all DX_i's should be "shrunk" or the process terminated if required accuracy has been achieved (all present DX_i's are within accuracy limits).
6. If X_1 is not the "center" point, then some improvement in the function has been found, and the algorithm attempts to take a larger step, moving to

$$X = X_1 + (X_1 - X_{base})$$

If the function is improved at X, then
(a) X becomes the new "center" for a local exploration.
(b) X_{base} takes on the value X_1.
If the function is not improved by moving to X, then X_{base} takes on the last "center" value.

The algorithm cycles through this sequence of steps until all DX_i's have been shrunk to some predetermined size that produces acceptable accuracy in the location of the minimum.

6.7 COMPARISONS OF SEARCH AND GRADIENT METHODS

The pattern search algorithm frequently has a real performance advantage over other algorithms, particularly the straightforward gradient algorithm (steepest descent). Some performance comparisons are in order. When researchers began proposing various different optimization algorithms it became clear that there is some advantage to having some standard problems for which performance could be compared. The most famous involves the Rosenbrock function [7]:

$$R(\mathbf{X}) = 100(X_2 - X_1^2)^2 + (1 - X_1)^2 \tag{6.7.1}$$

Traditionally, the search is started from $X_1 = -1.2$ and $X_2 = 1$. This particular function seems to be particularly nasty. See Daniels [7] for a sketch of the function and a discussion of the problems minimizing the Rosenbrock function.

More examples of troublesome functions are given in the miniprojects at the end of this chapter. However, we should note that there are many different kinds of problems, and no one method is best for all. However, there are times when we wish to modify a function in order to optimize under constraints, and frequently a search method, like pattern search, is better than a gradient method.

PROBLEM 6.7

Compare the performance of the pattern search and gradient method for finding the best model in Example 6.2. The best method is that one which uses the smallest number of function evaluations.

PROBLEM 6.8

Comparison of optimization algorithms is difficult because performance is often tied to the particular problem and starting point. We have discussed several optimization problems to this point, including

RCFILTER: Finding R's for best frequency response

TRANSFORMERWEIGHT: Finding dimensions of lightest transformer

Rosenbrock function

Model parameter identification (Problem 6.2)

We have also discussed a number of algorithms:

Gradient

Quadratic jump

Fletcher-Powell

Simplex

Pattern search (Hooke and Jeeves)

Choose one of these optimization problems and choose two of the algorithms and do a detailed study of the performance of the two algorithms on the problem, emphasizing the total number of calls on the FOFX subroutine in each case (as a measure of performance).

6.8 OPTIMIZATION UNDER CONSTRAINTS

Many optimization problems have some sort of constraint that is imposed on the parameters. For example, in the design of an *RC* filter, resistors and capacitors would have to be positive to be realistic. (Although it is possible to get negative resistors and capacitors using active circuits, we will not have negative resistors and capacitors if we are talking about passive circuits.) There are many other instances of constraints on parameters. Physical dimensions (lengths, for example) have to be positive. Frequently there are upper limits on sizes as well. There may be a limit on possible resistance, capacitance, or length in any given problem. Whatever the constraint, it must be taken into account in the solution of a problem or else the solution could be of no real use.

Constraints come in two general categories. Usually a constraint is either an equality constraint or an inequality constraint. An equality constraint requires equality of some function(s) of the parameters. For example,

$$X_1 + 2X_2^2 = 0 \qquad (6.8.1)$$

is an equality constraint establishing a relationship between the two parameters X_1 and X_2. On the other hand, an inequality constraint could be

$$X_1 + 2X_2^2 < 0 \qquad (6.8.2)$$

Generally, an equality constraint forces allowable points into a subspace of the parameter space (e.g., a curve in two dimensions). An inequality constraint restricts allowable points to a portion of the search space; see Figure 6.20, which shows the allowable regions for the two constraints just discussed.

We can account for constraints in several different ways. In this section we will discuss an analytical approach called the method of "slack variables." Later we will examine more heuristic numerical techniques.

Let us imagine that we have a two-dimensional optimization problem with variables X_1 and X_2. Further imagine that X_2 is constrained to be positive, so $X_2 > 0$. If we were to replace X_2 everywhere we found it with another variable, X_3^2, we would be replacing X_2 with a variable that *must* be positive.

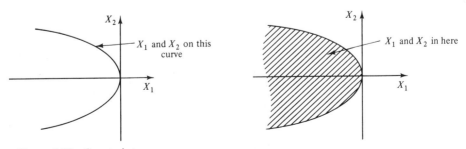

Figure 6.20 Constraints.

EXAMPLE 6.3

Optimize the function

$$f(\mathbf{X}) = 4(X_1 + 1)^2 + 5(X_2 + 2)^2$$

Subject to the constraint $X_2 \geq 0$.

Replace X_2 with a slack variable that is a square, and minimize the new function instead:

$$X_3^2 = X_2$$
$$f(\mathbf{X}) = 4(X_1 + 2)^2 + 5(X_3^2 + 2)^2$$

Whenever we must deal with an inequality constraint we can introduce a slack variable if we can put the constraint inequality into the form

$$C(\mathbf{X}) > 0 \qquad (6.8.3)$$

EXAMPLE 6.4

Minimize $(X_1 - 5)^2 + 10(X_2 - 7)^2$ subject to $X_1 + X_2 < 3$. Let

$$C(\mathbf{X}) = X_3^2 = (3 - X_1 - X_2) > 0$$

Then solve for X_2 in terms of X_3 and X_1, eliminating X_2 everywhere:

$$X_2 = 3 - X_1 - X_3^2$$

and hence

$$f(\mathbf{X}) = (X_1 - 5)^2 + 10(X_2 - 7)^2 = (X_1 - 5)^2 + 10(-X_1 - X_3^2 - 4)^2$$

This function can now be minimized using any method we choose, including any analytical or numerical method we might particularly favor.

PROBLEM 6.9

In Chapter 5, Section 5.9, we considered optimizing the frequency response of a simple RC filter. Modify the FORTRAN program that calculates the performance index to take into account a constraint that the resistors both be positive.

6.9 PENALTY FUNCTIONS FOR CONSTRAINED OPTIMIZATION

Constraints can also be accounted for numerically. However, since there can be either equality or inequality constraints, we might have to consider different methods for each type of constraint. The two types of constraints have to be handled differently, especially when considering analytical treatments. However, there is one numerical technique that works for both inequality and equality constraints, and we want to look at that method next.

One method of forcing the optimization algorithm to recognize constraints

is to add a large penalty to the performance index whenever the constraint is violated. An example will help clarify what we mean.

EXAMPLE 6.5

Consider the two-dimensional function

$$F(\mathbf{X}) = (X_1 - 5)^2 + 10(X_2 - 7)^2$$

We want to require that $X_1 + X_2 < 3$. In order to minimize $F(\mathbf{X})$ and not violate the constraint, we will use instead the function

$$F_G(\mathbf{X}) = (X_1 - 5^2) + 10(X_2 - 7)^2 + G(\mathbf{X})$$

where $G(\mathbf{X})$ takes on a large value whenever the constraint is violated (whenever $X_1 + X_2 \geq 3$). One way of implementing this is in the FORTRAN subroutine FOFX given in Table 6.2.

Finding the minimum of this particular function is not the easiest numerical task. Clearly, if we attempt to evaluate a gradient near the discontinuity at the constraint border, the gradient will become quite large; hence any simple gradient algorithm will take a large step away from the constraint boundary, even if the actual solution lies on the boundary (as it frequently does). A pattern search algorithm gets hung up on the boundary because it searches in 90° directions, and the constraint boundary line direction is apparently just right to cause a breakdown of the pattern search algorithm. A simplex approach seems to work for this particular problem. The details are left as a problem.

PROBLEM 6.10

Compare gradient, pattern, and simplex searches for the constrained function in Example 6.5, and verify that the simplex approach works best for this problem.

We have to be careful in drawing conclusions here. Although the simplex algorithm is best here, other problems with other constraints may not produce the same conclusions. Perhaps the only conclusion to draw from this section is

TABLE 6.2

```
C
C***************************************************************
C
      SUBROUTINE FOFX(X,N,F)
      DIMENSION X(10)
C
C EXAMPLE FUNCTION FOR PENALTY FUNCTION.
C
      F = ((X(1) - 2.)**2) + 10.*((X(2) - 2.)**2)
      IF ((X(1) + X(2)) .GE. 3) THEN
         FPENALTY = ((X(1) + X(2))-3)*1.E+6
         F = F + FPENALTY
      ENDIF
      RETURN
      END
```

that caution is warranted, and some numerical experimentation with different algorithms may make an unworkable problem solvable.

REFERENCES

1. B. D. Bunday, *Basic Optimization Methods,* Edward Arnold, Baltimore, 1984. (This is a very well-written, easy-to-read introduction to optimization methods. It contains many implementations of the various algorithms, all written in BASIC.)
2. R. L. Zahradnik, *Theory and Techniques of Optimization for Practicing Engineers,* Barnes and Noble, New York, 1971.
3. R. Fletcher and M. J. D. Powell, A rapidly convergent descent method for minimization, *Computer Journal,* July 1963, pp. 163–168.
4. S. S. Rao, *Optimization Theory and Applications,* 2nd ed., Halsted Press/Wiley, New York, 1984
5. G. S. G. Beveridge and R. S. Schechter, *Optimization: Theory and Practice,* McGraw-Hill, New York, 1979, pp. 383–392. (A comprehensive treatment of many topics in optimization. Well written and complete.)
6. P. E. Gill, W. Murray, and M. H. Wright, *Practical Optimization,* Academic Press, New York, 1981. (A very thorough treatment of the subject of optimization. This book should be on your bookshelf as a principal reference.)
7. R. W. Daniels, *An Introduction to Numerical Methods and Optimization Techniques,* North-Holland, New York, 1978, pp. 195–202.

MINIPROJECTS

1. The pattern search method takes a pattern move that is a linear extrapolation in a direction defined by two "base points." After local exploration, the most recent base point, X_b, and the previous base point, $\mathbf{X}_{b(\text{old})}$, define a direction to move. The new starting point for local exploration is $\mathbf{X}_b + (\mathbf{X}_b - \mathbf{X}_{b(\text{old})})$.

   It might be possible to improve the pattern move by taking more information into account. If we have $\mathbf{X}_b$ and $\mathbf{X}_{b(\text{old})}$, call the base point before that $\mathbf{X}_{b(\text{older})}$. Then, those three points are connected by a quadratic curve given by

   $$\mathbf{X} = \mathbf{X}_{b(\text{older})} - \alpha(\alpha - 2)(\mathbf{X}_{b(\text{old})} - \mathbf{X}_{b(\text{older})}) + \frac{\alpha(\alpha - 1)}{2}(\mathbf{X}_b - \mathbf{X}_{b(\text{older})})$$

   where α is a parameter. Plug in values of 0, 1, and 2 for α and $\mathbf{X}$ takes on the values $\mathbf{X}_{b(\text{older})}$, $\mathbf{X}_{b(\text{old})}$, and $\mathbf{X}_b$, respectively. Then if the new point for local exploration is taken as the $\mathbf{X}$ found for a value of 3 for the parameter, our rule for computing the new point for local exploration is

   $$\mathbf{X}_{\text{new}} = \mathbf{X}_{b(\text{older})} - 3\mathbf{X}_{b(\text{old})} + 3\mathbf{X}_b$$

   Investigate the properties of this algorithm when applied to the Rosenbrock function, comparing convergence rates.

2. Assume that you know that a system has a DC gain of unity, and a single corner frequency. Assume also that you have available some experimental frequency response data on the system. The problem is to determine the best corner frequency to match the experimental data.

This problem can be posed as a univariate search problem if we can define a performance index.

(a) Write and check out a program using the Golden Section search procedure using a sum-of-squares-of-errors criterion. Redo for a sum-of-absolute-errors criterion.

(b) Redo part (a) using decibels instead of straight gain.

3. Factorizing a polynomial, $P(s)$, can be posed as a problem of minimizing $|P(s)|$ at some point in the complex s plane.

(a) Write a program that uses pattern search to minimize $|P(s)|$, letting X_1 and X_2 be the real and imaginary parts of s.

(b) Modify the program of part (a) to divide out the found root and eventually find all roots. You may use programs from Chapter 9.

4. Frequently the frequency response of a system has a resonant peak. It may be of importance to determine the frequency of resonance and the magnitude of the peak (usually expressed in decibels).

Assume that you have a transfer function, and that the coefficients of the numerator and denominator polynomials are stored in arrays PN and PD, respectively. The constant term is in PN(1), the coefficient of s in PN(2), and so forth. Write a program that asks the user for search limits, and then does a Golden Section search between those limits to determine the location of the peak. Note that the techniques discussed in this chapter assume a minimization problem, but that conversion to decibels followed by a sign change will change this maximization problem to a minimization problem.

5. We have considered finding the parameters of a dynamic model in Problem 6.3 (a single-time-constant problem) and in Example 6.2 (finding a time constant and a DC gain in a model). A slightly more complex model has two time constants and a DC gain, as shown in the figure below.

Input unit step $\longrightarrow$ $\boxed{\dfrac{a\,b\,G}{(s+a)(s+b)}}$ $\longrightarrow$ Response $r(t)$

The step response of this system takes two forms, depending upon whether or not the two time constants are equal:

$$r(t) = \left[1 - \frac{be^{-at}}{b-a} + \frac{ae^{-bt}}{b-a}\right] \qquad a \neq b$$
$$r(t) = [1 - e^{-at} - ate^{-at}] \qquad a = b$$

This "singular case" poses problems if we try to use these two forms to compare to some measured data, and do a parameter fit (as we have done in Problem 6.3 and Example 6.2). When the two time constants are equal, then there is a sensitivity problem, since the two exponential portions of the response are subtracted, and are nearly equal.

One possible way around this difficulty is to do a numerical computation of the response of the system using a standard integration algorithm and comparing the simulation results with the measured response of the system. At the end of this problem is a set of data for such a system.

(a) Draw a flowchart of a method for using a simulation to compute the response and compare that computed response to the measured response so that a minimization problem similar to Example 6.3 can be set up.

(b) Program and test your algorithm. (Multiply sample number by 5.0 to get time.)

```
SAMPLE
NUMBER      RESPONSE
5.000000    22.55900
10.00000    29.47600
15.00000    38.06700
20.00000    46.29800
25.00000    53.76800
30.00000    60.45200
35.00000    66.25500
40.00000    71.44800
45.00000    76.03300
50.00000    80.19400
55.00000    83.95200
60.00000    87.39900
70.00000    93.47200
80.00000    98.70500
90.00000    103.1960
100.0000    106.9980
110.0000    110.3220
120.0000    113.1600
130.0000    115.8160
140.0000    118.2530
154.0000    120.8820
169.0000    123.2300
188.0000    125.9720
213.0000    129.0700
238.0000    131.1250
```

Chapter 7

Generating State Equations from Topological Data

7.1 TOPOLOGICAL CONCEPTS: AN INTRODUCTION

In this chapter we will investigate generating state equation descriptions of networks starting from a basic topological description of a circuit. Our goal here is to develop techniques for taking a basic topological description (that is, information about how elements are interconnected) and generating the kinds of information that are useful in calculating things like transient response, transfer functions, and the like. The goal of our machinations is a set of state equations with numerical values for all of the matrices involved in the set of state equations. We will be fairly informal about this process. There are several texts and papers that present this sort of material, with varying degrees of rigor and readability [1–4]. In this chapter we will examine only linear circuits. Nonlinear circuits cannot be ignored, however. Although we will not cover nonlinear circuits, the interested reader will find a good treatment of the subject in Chua and Lin [4].

Our first goal will be to obtain a description of the topology of an electrical network—that is, some sort of mathematical description of the interconnections of all the elements in the network. Then we will take that description and generate the kind of topological information usually used by humans in the form of Kirchhoff's current and voltage laws. Finally, we will combine that topological information with mathematical descriptions of the elements forming the network. We will finish with a set of state equations in standard form so that we will be ready to use those state equations for further analysis, such as finding the transfer function and doing transient analysis using matrix exponential techniques.

All of the information about interconnections in a circuit can be summarized in an *Incidence matrix*. To introduce this concept, let us examine a simple example network, the one shown in Figure 7.1, and see how the incidence

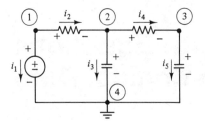

Figure 7.1 An example circuit.

matrix can be used to code topological information. If we write Kirchhoff's current law at nodes 1, 2, 3, and 4, we get the following set of equations:

$$
\begin{array}{llll}
\text{Node 1:} & i_1 + i_2 & = 0 & \text{(7.1.1a)} \\
\text{Node 2:} & -i_2 + i_3 + i_4 & = 0 & \text{(7.1.1b)} \\
\text{Node 3:} & -i_4 + i_5 = 0 & & \text{(7.1.1c)} \\
\text{Node 4:} & -i_1 \quad - i_3 \quad - i_5 = 0 & & \text{(7.1.1d)}
\end{array}
$$

One thing we can see immediately is that there is redundancy in this set of equations, since Eq. (7.1.1d) can be obtained from the other three equations (7.1.1a–c) by adding all three together. Usually, in a case like this we will ignore the Kirchhoff current law equation for the ground node, and use the remaining equations. With the ground node equation in the set of equations, we can see that every current appears exactly twice, once added, and once subtracted. If we ignore the redundant information in the ground node equation, the information in these equations can be expressed compactly in vector matrix format:

$$
\begin{bmatrix}
1 & 1 & 0 & 0 & 0 \\
0 & -1 & 1 & 1 & 0 \\
0 & 0 & 0 & -1 & 1
\end{bmatrix}
\begin{bmatrix}
i_1 \\ i_2 \\ i_3 \\ i_4 \\ i_5
\end{bmatrix}
=
\begin{bmatrix}
0 \\ 0 \\ 0
\end{bmatrix}
\tag{7.1.2}
$$

This may be more compactly written [1] as

$$
A * i = 0 \tag{7.1.3}
$$

where

$$
A =
\begin{bmatrix}
1 & 1 & 0 & 0 & 0 \\
0 & -1 & 1 & 1 & 0 \\
0 & 0 & 0 & -1 & 1
\end{bmatrix}
\qquad
i =
\begin{bmatrix}
i_1 \\ i_2 \\ i_3 \\ i_4 \\ i_5
\end{bmatrix}
\tag{7.1.4}
$$

Equation (7.1.3) neatly summarizes in vector matrix form all of the information contained in the Kirchhoff current law equations. The matrix A is usually referred to as the incidence matrix.

PROBLEM 7.1

Find the incidence matrix for the two networks shown.

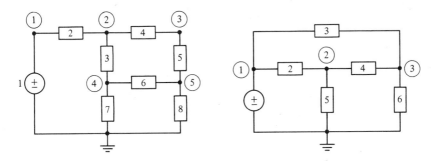

PROBLEM 7.2

Construct the incidence matrix for the network shown. The two sections of the network are independent. Determine what happens to the incidence matrix if they are completely disconnected. Is it really necessary to refer each section to ground?

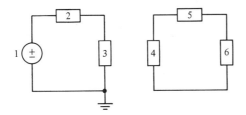

However, the current law equations do not use effectively all of the information about the structure of the network. The structure of any electrical network defines both a set of current law equations and a set of voltage law equations. The current and voltage law equations are half of what we need to consider. The other half of the information we need is contained in the relations that describe the elements—what they are and what their numerical values are. For now, however, the current law equations are a starting point and we will work from our incidence matrix description toward a set of network state equations.

The incidence matrix formulation of Kirchhoff's current law can be manipulated to provide more information about the circuit. Eventually we want to be able to determine which elements constitute loops in the circuit—for example, whenever we want to deal with Kirchhoff's voltage law formulations. If we can manipulate this expression so that the incidence matrix can be partitioned into an identity matrix and another matrix, then later we will be able to manipulate the circuit equations much more easily, particularly when we try to generate state equations numerically.

Let us pursue this approach a little further by reducing this particular incidence matrix to the form we want. Normally, this reduction would be done using Gaussian reduction, but for this example we can simply rearrange the order of the elements. The order we will choose is 1, 3, 5, 2, 4. Rewriting the Kirchhoff current law equations in the incidence matrix format, we get

$$\begin{bmatrix} 1 & 0 & 0 & : & 1 & 0 \\ 0 & 1 & 0 & : & -1 & 1 \\ 0 & 0 & 1 & : & 0 & -1 \end{bmatrix} \begin{bmatrix} i_1 \\ i_3 \\ i_5 \\ i_2 \\ i_4 \end{bmatrix} = 0 \qquad (7.1.5)$$

This is the form we want:

$$[I : F] * i = 0 \qquad (7.1.6)$$

Once we have KCL in this peculiar form, we can notice certain circuit implications. In the circuit, each KCL equation in this formulation involves exactly one current from the set (i_1, i_3, i_5). That set of elements, (1, 3, 5), connects each node in the circuit, but contains no loops. Thus, the set of elements (1, 3, 5) forms a *tree* of the network. This tree is shown in Figure 7.2. Each KCL equation in this formulation involves precisely one tree current, and various other *link* currents.

PROBLEM 7.3

For the circuits in Problems 7.1 and 7.2 do a Gaussian reduction to find trees for each circuit. Sketch the tree for each circuit.

7.2 KIRCHHOFF'S LAWS FROM TOPOLOGICAL DATA

Before we can go much further, we need to define our vocabulary more clearly. Let us accept the following as working definitions of our terms (see Desoer and Kuh [1]):

Branch. An idealization of a two-terminal electrical element—that is, one with two external leads.

Node. An idealization of a point in space where leads of electrical elements are connected together.

Graph. An idealization of an electric circuit. A graph is composed of branches connected together at nodes.

Path. A set of branches that connects two nodes but has no node appearing twice.

Loop. A closed path—that is, a path in which the starting node and ending node are the same.

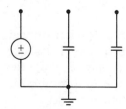

Figure 7.2 A network tree.

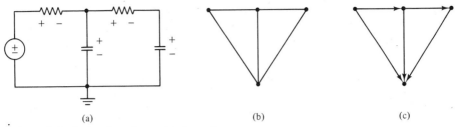

(a)   (b)   (c)

Figure 7.3 A circuit and associated graphs.

For any given electric circuit composed of two terminal elements, we can construct a graph representing the circuit. In Figure 7.3 we see a circuit (a), and the graph we construct for the circuit (b). This type of graph does not completely present all of the information we might need to analyze the circuit. In particular, when dealing with electrical elements we need to know something about the polarity of the voltage or current. In Figure 7.4 we see a general element with assumed directions for the voltage across the elements and the current through the element. This set of polarities is usually termed the "load reference" set of polarities. (With load reference polarities, power into the element load is VI.) We will always assume that our references are of this sort, and we will associate a direction with each element to indicate polarity. That direction is shown in Figure 7.4 for the single element, and is incorporated into Figure 7.3c, where it indicates all polarities associated with voltages and currents in the circuit of Figure 7.3a. A graph of this sort is called a *directed graph,* but we will usually neglect to make this distinction in our presentation.

For any directed graph, define an incidence matrix to summarize all of the interconnection information in the graph. Define a matrix A with elements A_{ij} such that

$$A_{ij} = +1 \text{ if branch } j \text{ leaves node } i \qquad (7.2.1a)$$
$$= -1 \text{ if branch } j \text{ enters node } i \qquad (7.2.1b)$$
$$= 0 \text{ if branch } j \text{ is not connected to node } i \qquad (7.2.1c)$$

For our circuit in Figure 7.3 we obtain the incidence matrix A as

$$A = \begin{bmatrix} 1 & 1 & 0 & 0 & 0 \\ 0 & -1 & 1 & 1 & 0 \\ 0 & 0 & 0 & -1 & 1 \\ -1 & 0 & -1 & 0 & -1 \end{bmatrix} \qquad (7.2.2)$$

One interesting property of the incidence matrix is immediately apparent. All columns of the incidence matrix have exactly one "1" entry and one "−1"

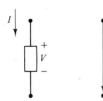

Figure 7.4 An isolated element and associated graph branch.

entry. All other entries are zero. Any single row can be eliminated without loss of information. (If necessary, we could later regenerate the row we removed.) In electric circuits, removal of one row corresponds to picking one node as the ground or reference or datum node. If a row is eliminated, then the resulting matrix is called a "reduced incidence matrix," although, once again, we will not usually make that distinction in our presentation here.

Delete the fourth row (saying in the process that node 4 is the ground node) to get

$$A = \begin{bmatrix} 1 & 1 & 0 & 0 & 0 \\ 0 & -1 & 1 & 1 & 0 \\ 0 & 0 & 0 & -1 & 1 \end{bmatrix} \tag{7.2.3}$$

This is the same matrix encountered in the introductory example.

Clearly, Kirchhoff's current law equations are contained in the information coded into the incidence matrix. What is not as clear is that some information about voltages is also hidden there. For instance, we can define two voltage vectors as follows:

$$\mathbf{V}_{branch} = \text{Vector of branch voltages} \tag{7.2.4a}$$
$$\mathbf{V}_{node} = \text{Vector of node voltages} \tag{7.2.4b}$$

In both cases we assume that subscripts come in numerical order. $\mathbf{V}_{branch}$ has branch voltages in numerical order, and $\mathbf{V}_{node}$ has node voltages in numerical order. Now, any branch voltage is the difference of two node voltages. Figure 7.5 shows the relationship between node voltages and branch voltages for a typical element. This implies that the vector of branch voltages, $\mathbf{V}_{branch}$, is linearly related to the vector of node voltages, $\mathbf{V}_{node}$, in a vector matrix relation as follows:

$$\mathbf{V}_{branch} = C * \mathbf{V}_{node} \tag{7.2.5}$$

Here, the entries in the matrix, C, are defined as follows:

$$C_{ji} = +1 \text{ if "+" end of branch } j \text{ is connected to node } i \tag{7.2.6a}$$
$$= -1 \text{ if "--" end of branch } j \text{ is connected to node } i \tag{7.2.6b}$$
$$= 0 \text{ if branch } j \text{ is not connected to node } i \tag{7.2.6c}$$

However, we see that we are really defining a matrix C that is equal to the transpose of the incidence matrix A, which was previously defined. What we really have is

$$\mathbf{V}_{branch} = A^T * \mathbf{V}_{node} \tag{7.2.7}$$

We see that the incidence matrix not only contains information about Kirchhoff current law equations but also can be used to relate branch voltages

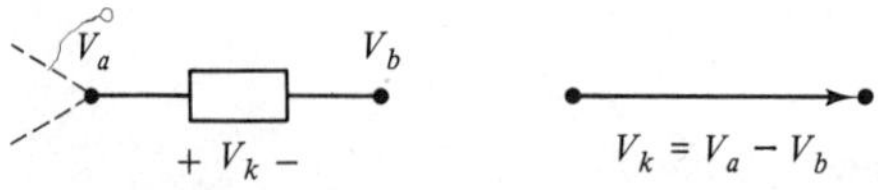

Figure 7.5 Voltage conventions.

and node voltages. For a more complete discussion see Desoer and Kuh [1] or Chua and Lin [4].

7.3 A MORE GENERAL FORMULATION FOR RESISTIVE NETWORKS

The graph of a network condenses all of the topological information about a network into a concise, visual model. However, the graph of a network contains no information about the details of the elements in the network. Normally, we would assume that the network consisted of voltage and current sources, resistors, capacitors, inductors, transistors, and the like, with each element composed of one "elementary" element. In this section, we will take a short detour and examine some resistive networks composed of composite elements. This approach, which will contrast with the conductance matrix approach developed in Chapter 1, is sometimes used instead of the conductance matrix model.

We can use the information contained in the incidence matrix to obtain a more general solution of the circuit equations for resistive networks. We will work with a general branch, of the form shown in Figure 7.6. This branch contains a conductance (resistance), a current source, and a voltage source. With these components we can represent a variety of other types of branches. A resistance, a voltage source with internal resistance, and a current source with internal resistance can all be considered to be special cases of this kind of branch.

We can write the general equation for a branch as

$$V_k = R_k(J_k - J_{sk}) + V_{sk} \tag{7.3.1}$$

or

$$I_k = G_k(V_k - V_{sk}) + I_{sk} \tag{7.3.2}$$

If we write things in terms of vectors of branch voltages and currents, we have

$$\mathbf{I}_{branch} = G * \mathbf{V}_{branch} + \mathbf{J}_s - G * \mathbf{V}_{bs} \tag{7.3.3}$$

where

$$\mathbf{I}_{branch} = \text{Vector of branch currents} \tag{7.3.4a}$$
$$\mathbf{V}_{branch} = \text{Vector of branch voltages} \tag{7.3.4b}$$
$$\mathbf{J}_s = \text{Vector of branch current sources} \tag{7.3.4c}$$

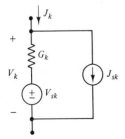

Figure 7.6 A general resistive branch.

$\mathbf{V}_s$ = Vector of branch voltage sources   (7.3.4d)

G = diagonal matrix of conductances

$(G_{ii} = G_i = i$th conductance)   (7.3.4e)

Now, if we premultiply the vector of branch currents by the incidence matrix, we are setting up Kirchhoff's current law equations.

$$A * \mathbf{I}_{\text{branch}} = A * (G * \mathbf{V}_{\text{branch}} + \mathbf{J}_s - G * \mathbf{V}_{bs}) = 0 \qquad (7.3.5)$$

Then, using the relation between $\mathbf{V}_{\text{branch}}$ and $\mathbf{V}_{\text{nodes}}$, we find

$$A * G * A' * \mathbf{V}_{\text{nodes}} = -A * \mathbf{J}_s + A * G * \mathbf{V}_{bs} \qquad (7.3.6)$$

We can now solve for the node voltages as

$$\mathbf{V}_{\text{nodes}} = (A * G * A')^{-1} * (-A * \mathbf{J}_s + A * G * \mathbf{V}_{bs}) \qquad (7.3.7)$$

PROBLEM 7.4

Set up the incidence matrix, conductance matrix, and source vector for the networks shown.

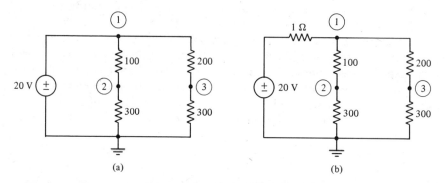

(a)   (b)

This formulation using generalized elements has been used occasionally as a basis for circuit analysis programs. However, more general formulations are more useful, particularly a formulation that does not require every source to be accompanied by an internal resistance. Even though every source really has some internal resistance, it is often best if we can dispense with internal resistance particularly in voltage sources.

7.4 FINDING A TREE

The incidence matrix allows us to formulate Kirchhoff's current law from a topological description. However, the ultimate goal is to generate a set of state equations. Kirchhoff law equations are only an intermediate step in the complete analysis. To get state equations requires bringing together information from Kirchhoff's current law and Kirchhoff's voltage law, as well as information describing the voltage-current characteristic of each individual element. The critical concept we need to develop is that of a tree in a circuit, and numerical methods for finding trees in circuits. First, we give a definition of what we mean by a tree in a circuit.

Tree. A subset of the branches of a network that
1. Connects all of the nodes within a network.
2. Contains no loops.

Figure 7.7 shows a network, and three of the possible trees of this network. It should be clear from this that any given network can have many trees. Larger networks can have incredibly large numbers of possible trees.

PROBLEM 7.5
Find and sketch all of the trees in Figure 7.8.

However, while a network can have a number of trees, each tree has the same number of elements. If the number of branches in a network is N_{branches}, and the number of nodes is N_{nodes}, then we will have $(N_{\text{nodes}} - 1)$ branches in any given tree, and $(N_{\text{branches}} - N_{\text{nodes}} + 1)$ branches not in the tree. To see this, consider the first branch in a tree. That branch connects two nodes, and each further branch connects just one more node, so when all nodes are connected, the number of branches used is $(N_{\text{nodes}} - 1)$. (If a branch does not connect a new node to the network, it must "double back" to a node already connected, creating a loop in the process.) The branches not in the tree are called link branches or just "links."

Now we continue examining how we might find a tree numerically and what the implications are for the incidence matrix when we have identified a tree. In Section 7.1 we pointed out that, for the example given there, we could manually identify a tree and then rearrange elements so that the tree branch currents appeared first in the list. When we did this, we were able to put the incidence matrix in the form

$$A = [I \;\vdots\; F] \tag{7.4.1}$$

We need to identify an algorithm that will produce this form of incidence matrix so that we don't have to do it manually. One possible algorithm is an adaption

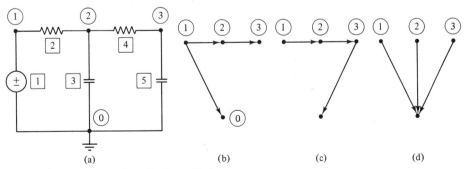

Figure 7.7 A network and a few of its trees.

of Gaussian elimination. Let us go back to the original incidence matrix we obtained for the circuit we had in Figure 7.1. That incidence matrix is

$$A = \begin{bmatrix} 1 & 1 & 0 & 0 & 0 \\ 0 & -1 & 1 & 1 & 0 \\ 0 & 0 & 0 & -1 & 1 \end{bmatrix} \qquad (7.4.2)$$

Now start with a Gaussian elimination procedure. The first column of this matrix already has a one on the diagonal and zeros elsewhere, so we don't have to do anything with that column. Our first real operation on the matrix will be to change sign on the second row, and then subtract the second row from the first row, yielding

$$A_1 = \begin{bmatrix} 1 & 1 & 0 & 0 & 0 \\ 0 & 1 & -1 & -1 & 0 \\ 0 & 0 & 0 & -1 & 1 \end{bmatrix} \quad \text{(Changing signs in row 2)} \qquad (7.4.3)$$

and $\qquad A_2 = \begin{bmatrix} 1 & 0 & 1 & 1 & 0 \\ 0 & 1 & -1 & -1 & 0 \\ 0 & 0 & 0 & -1 & 1 \end{bmatrix} \qquad (7.4.4)$

When we get to this point, we can see problems have developed. There is a zero on the diagonal at the (3, 3) location, and there is no way to proceed further in the Gaussian reduction using only the first three elements. What is happening here? The inability to complete the Gaussian reduction is an indication that the elements we are working with contain a loop. If we look at the circuit's graph in Figure 7.8, it is clear that elements 1, 2, and 3 cannot form a tree, since in fact they form a loop.

In order to continue the Gaussian elimination, the only option we have is to interchange another column with the third column, so that we move a nonzero element into the (3, 3) position. We have a choice, since we can switch in either the fourth or the fifth element (by interchanging column 3 with either column 4 or column 5). The two possibilities yield either (1, 2, 4) or (1, 2, 5) as the set of three elements, and each of these set of branches constitutes a tree (see Figure 7.9). If we arbitrarily say that branch 4 is switched, we have

$$A_3 = \begin{bmatrix} 1 & 0 & 1 & 1 & 0 \\ 0 & 1 & -1 & -1 & 0 \\ 0 & 0 & -1 & 0 & 1 \end{bmatrix} \qquad (7.4.5)$$

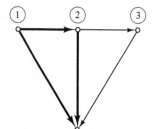

Figure 7.8 Emphasizing a set of elements comprising a loop.

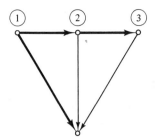

Figure 7.9 Emphasizing a set of elements comprising a tree.

Completing the Gaussian reduction procedure, we have

$$A_4 = \begin{bmatrix} 1 & 0 & 1 & 1 & 0 \\ 0 & 1 & -1 & -1 & 0 \\ 0 & 0 & 1 & 0 & -1 \end{bmatrix} \qquad (7.4.6)$$

$$A_5 = \begin{bmatrix} 1 & 0 & 1 & 1 & 0 \\ 0 & 1 & 0 & -1 & -1 \\ 0 & 0 & 1 & 0 & -1 \end{bmatrix} \qquad (7.4.7)$$

$$A_6 = \begin{bmatrix} 1 & 0 & 0 & 1 & 1 \\ 0 & 1 & 0 & -1 & -1 \\ 0 & 0 & 1 & 0 & -1 \end{bmatrix} \qquad (7.4.8)$$

We need to be alert to the fact that the order of branches is now (1, 2, 4, 3, 5), corresponding to the columns in the incidence matrix. (And we have to keep track of branch order in any program we write!) In any event, in the incidence matrix, A_6, which has the identity matrix in the left square, we find some interesting relations. The Kirchhoff current law equations we get using this incidence matrix are

$$i_1 \quad + i_3 + i_5 = 0 \qquad (7.4.9a)$$
$$i_2 \quad - i_3 - i_5 = 0 \qquad (7.4.9b)$$
$$i_4 \quad - i_5 = 0 \qquad (7.4.9c)$$

Each of these equations is the equation for Kirchhoff's current law for a particular *cut set* of branches. In Figure 7.10, we can see the cut set to which each equation applies. We will define a cut set as follows:

> *Cut set.* A set of branches in a graph that when removed, separates the graph into two disjoint subgraphs.

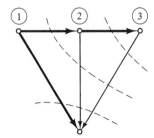

Figure 7.10 Three cut sets.

Cut set current equations are a generalization of Kirchhoff's current law equations. Each cut set of branches contains exactly one branch of the tree being used, plus as many links as are necessary to separate the graph.

When a graph has been separated into two parts in this manner, we can argue physically why the sum of currents is zero through the *Gaussian surface* separating the two parts of the circuit. There is no net charge accumulation on any element in the circuit, nor is there any net charge accumulation at nodes. (Even in a capacitor, which stores charge, the *net* charge on the capacitor will always be zero, with equal amounts of positive and negative charge on the two capacitor plates.) Thus the net charge flow across the surface separating the two parts of a circuit must be zero, and since the charge flows through the leads between the two parts of the circuit, the net current flowing between any two subcircuits separated by a Gaussian surface will be zero. Each cut set defines a family of such surfaces, and the sum of the currents flowing through the cutset is zero.

PROBLEM 7.6

It is usually desirable to determine a tree in which all of the voltage sources and capacitors appear.

(a) In the circuit shown, how many trees contain both the voltage source and the capacitor? Sketch them.

(b) It is also desirable to exclude current sources and inductors from the tree. How many trees exist in the circuit if the current source is excluded but the voltage source and capacitor are included?

(c) For each of the trees found in part (b), write the current law equations in incidence matrix form.

(d) For each tree found in part (b), identify all cut sets. Is each cut set associated with a unique current law equation in part (c)?

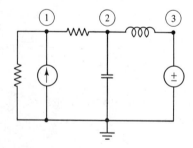

ALGORITHM

FINDING A TREE OF AN ELECTRIC CIRCUIT

1. Start with an arbitrary incidence matrix for a circuit.
2. Begin doing a Gaussian reduction, trying to form a matrix with ones on the diagonal and zeros below the diagonal.
3. If, at any point, it becomes impossible to continue, then interchange

columns to get a nonzero entry on the diagonal and continue the process of reduction.

4. With ones on the diagonal, and zeros below, eliminate all nonzero elements above the diagonal, to produce the form

$$[I \vdots F].$$

In applying this algorithm we will have to keep track of element order whenever columns are interchanged. So will a digital computer program. Once the algorithm is finished, the branches corresponding to the first $(N_{\text{nodes}} - 1)$ entries are branches of one possible tree. (Starting with a different element order will probably produce a different tree.)

7.5 VOLTAGE RELATIONS

The matrix F found in the preceding section also relates the voltages in the network. Take one of the trees found for our network. If we imagine adding one of the links to the tree, we form a loop, as in Figure 7.11. If we instead add the other link, we get a different loop. Writing Kirchhoff's voltage law for these two loops, we find

$$-v_1 + v_2 \quad + v_3 \quad = 0 \qquad (7.5.1a)$$

and
$$-v_1 + v_2 + v_4 \quad + v_5 = 0 \qquad (7.5.1b)$$

Putting this into vector matrix format, we get

$$\begin{bmatrix} -1 & 1 & 0 & 1 & 0 \\ -1 & 1 & 1 & 0 & 1 \end{bmatrix} \begin{bmatrix} v_1 \\ v_2 \\ v_4 \\ v_3 \\ v_5 \end{bmatrix} = 0 \qquad (7.5.2)$$

This is of the form

$$[G \vdots I] * \mathbf{V}_{\text{branch}} = 0 \qquad (7.5.3)$$

However, we know that the branch voltages are related to the node voltages since, from Eq. (7.2.7), $\mathbf{V}_{\text{branch}} = A^{\text{T}} * \mathbf{V}_{\text{node}}$, so we get

$$[G \vdots I] * \mathbf{V}_{\text{branch}} = [G \vdots I] * A^{\text{T}} * \mathbf{V}_{\text{node}} = 0 \qquad (7.5.4)$$

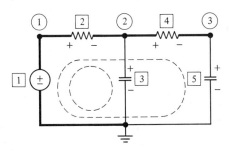

Figure 7.11 A tree and KVL.

Now if $A^T = [I \vdots F]$, then we have

$$[G \vdots I]*[I \vdots F]^T * \mathbf{V}_{node} = 0 \tag{7.5.5}$$

If we imagine a general network, with element types not specified, then any node voltages whatsoever could occur. Since (7.2.5) must hold for any $\mathbf{V}_{node}$ whatsoever, we must have

$$[G \vdots I]*[I \vdots F]^T = 0 \tag{7.5.6}$$
or
$$G + F^T = 0 \tag{7.5.7}$$
so
$$G = -F^T \tag{7.5.8}$$

In other words, the matrix F previously found for the circuit equations also gives us the Kirchhoff voltage law equations. (For a thorough, but readable, derivation of KVL using incidence matrix concepts see Desoer and Kuh [1].)

PROBLEM 7.7

Refer back to the circuit shown in Figure 7.1 of this chapter. From the incidence matrix:

(a) Generate the matrix form of Kirchhoff's voltage law.
(b) Sketch the circuit and the loops to which KVL applies, for the relations defined in part (a).

PROBLEM 7.8

The figure below shows a bridge circuit configuration. Using the element numbers and node labels as given:

(a) Construct the reduced incidence matrix for this bridge circuit.
(b) Find the "F matrix" for the circuit.
(c) Using the F matrix, construct the Kirchhoff voltage law equations, find the loops they correspond to, and verify the equations.

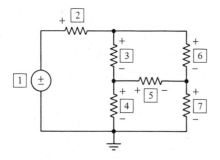

7.6 PROGRAMS TO IMPLEMENT THE
TREE-FINDING ALGORITHM

Programs to implement this kind of algorithm should do at least the following:

1. Provide input topological data (prompting as necessary).
2. Perform the Gaussian reduction.
3. Keep track of element order.

Programs that do this are given in Appendix H. The tasks are broken into several smaller tasks to keep program size small. The tasks done by programs in Appendix H are as follows:

1. Sort the circuit elements so that they are arranged in the order of voltage sources, capacitors, resistors, inductors, current sources (in CKTSRT).
2. Find a proper tree, generating an incidence matrix in the form $[I \vdots F]$ (in CKTREE).

7.7 GETTING THE NUMERICAL STATE EQUATIONS: AN EXAMPLE NETWORK

We are now finally at the point where we can begin getting the state equations of a network. We will work on the simple network we have been beating to death; then, in the next section we will generalize our techniques to other networks. This example will serve to illustrate the general approach that we will develop later. Thus far, we have generated a great deal of topological information, but the topological information isn't everything we need to know to generate state equations for a network. Sooner or later, we need to know elements types and values and get that information "cranked into" our analysis. This has to be done with some care. We notice that some resistors could be in the tree, and some could be in the link elements. These two cases have to be handled differently in our analysis.

To start that analysis let us define a set of vectors as follows. We will assume that we are dealing with a tree that contains the voltage source and the two capacitors, leaving the resistors to form the links. Why we do this will be obvious or will be clarified as we go on. The tree we are using is shown in Figure 7.12.

$$\mathbf{V}_{Tc} = \begin{bmatrix} v_3 \\ v_5 \end{bmatrix} = \text{Vector of tree capacitor voltages} \qquad (7.7.1a)$$

$$\mathbf{I}_{Tc} = \begin{bmatrix} i_3 \\ i_5 \end{bmatrix} = \text{Vector of tree capacitor currents} \qquad (7.7.1b)$$

$$\mathbf{V}_{Lg} = \begin{bmatrix} v_2 \\ v_4 \end{bmatrix} = \text{Vector of link resistor voltages} \qquad (7.7.1c)$$

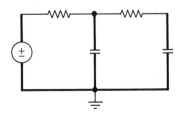

Figure 7.12 An example network.

$$\mathbf{I}_{Lg} = \begin{bmatrix} i_2 \\ i_4 \end{bmatrix} = \text{Vector of link resistor currents} \qquad (7.7.1d)$$

$$C_1 = \begin{bmatrix} C_1 & 0 \\ 0 & C_2 \end{bmatrix} \qquad (7.7.2a)$$

$$G = \begin{bmatrix} 1/R_1 & 0 \\ 0 & 1/R_2 \end{bmatrix} \qquad (7.7.2b)$$

$$\mathbf{E} = [v_1] \qquad (7.7.2c)$$

These vectors pull together all of the variables in our network.

The "constituent relationships" for the components may be written in vector matrix format as

$$\mathbf{I}_{Lg} = G * \mathbf{V}_{Lg} \qquad (7.7.3)$$

$$\mathbf{I}_{Tc} = C_1 * \frac{d(\mathbf{V}_{Tc})}{dt} \qquad (7.7.4)$$

The last equation here is the precursor of the state equation we want to get. We will work to transform that expression to get the state equations we need.

We will choose states to be the capacitor voltages. Since there are two energy storage elements in the network, we expect there are two states. Charges might be chosen charge, but capacitor voltages are also relatively natural choices for states.

Go back now to the incidence matrix. For our choice of tree branches we have

$$A = \begin{bmatrix} 1 & 0 & 0 & 1 & 0 \\ 0 & 1 & 0 & -1 & 1 \\ 0 & 0 & 1 & 0 & -1 \end{bmatrix} \qquad (7.7.5)$$

Once again, this is the form $[I \;\vdots\; F]$, and we can view this incidence matrix as setting up a relationship between tree variables and link variables. Define

$$\mathbf{V}_{\text{tree}} = \begin{bmatrix} \mathbf{E} \\ \mathbf{V}_{Tc} \end{bmatrix} \qquad \mathbf{V}_{\text{link}} = [\mathbf{V}_{Lg}] \qquad (7.7.6a,b)$$

and

$$\mathbf{I}_{\text{tree}} = \begin{bmatrix} \mathbf{I}_e \\ \mathbf{I}_{Tc} \end{bmatrix} \qquad \mathbf{I}_{\text{link}} = [\mathbf{I}_{Lg}] \qquad (7.7.6c,d)$$

and we will have

$$[I \;\vdots\; F] * \begin{bmatrix} \mathbf{I}_{\text{tree}} \\ \mathbf{I}_{\text{link}} \end{bmatrix} = 0 \qquad (7.7.7)$$

Solving for $\mathbf{I}_{\text{tree}}$,

$$\mathbf{I}_{\text{tree}} = -F * \mathbf{I}_{\text{link}} \qquad (7.7.8)$$

This last expression will always be true. We can now break down the tree current vector into current in the source and capacitor currents, and partition the matrix F in a corresponding manner, so we have

$$\mathbf{I}_{\text{tree}} = \begin{bmatrix} I_e \\ I_{Tc} \end{bmatrix} = \begin{bmatrix} -F_{eg} \\ -F_{cg} \end{bmatrix} * \mathbf{I}_{Lg} = -F * \mathbf{I}_{\text{link}} \qquad (7.7.9)$$

Now, if we solve for the numerical version, we can get the capacitor currents as:

$$\mathbf{I}_{Tc} = \begin{bmatrix} i_3 \\ i_5 \end{bmatrix} = \begin{bmatrix} 1 & -1 \\ 0 & 1 \end{bmatrix} \begin{bmatrix} i_2 \\ i_4 \end{bmatrix} \qquad (7.7.10)$$

or
$$\mathbf{I}_{Tc} = F_{cg} * \mathbf{I}_{Lg} \qquad (7.7.11)$$

This gives us the capacitor currents in terms of the link resistor currents. In turn, those currents are related to the link resistor voltages by

$$\mathbf{I}_{Lg} = G * \mathbf{V}_{Lg} \qquad (7.7.12)$$

Using this in the expression for capacitor currents gives:

$$\mathbf{I}_{Tc} = F_{cg} * G * \mathbf{V}_{Lg} \qquad (7.7.13)$$

Now, we need to express the link conductance voltages, $\mathbf{V}_{Lg}$ in terms of the states $\mathbf{V}_{Tc}$ and the input $\mathbf{E}$. We can get the relationship we need from the matrix Kirchhoff voltage law given by Eq. (7.5.3).

The KVL equations can be written in an expanded form, similar to the expanded form for the current equations:

$$[-F^{\mathrm{T}} \;\vdots\; I] * \mathbf{V}_{\text{branches}} = 0 \qquad (7.7.14)$$

or
$$[-F^{\mathrm{T}} \;\vdots\; I] * \begin{bmatrix} \mathbf{V}_{\text{tree}} \\ \mathbf{V}_{\text{link}} \end{bmatrix} = 0 \qquad (7.7.15)$$

If we break this up so that the tree voltage vector is expressed in terms of subvectors, we have

$$\mathbf{V}_{Lg} = \mathbf{V}_{\text{link}} + F^{\mathrm{T}} * \mathbf{V}_{\text{tree}} = \begin{bmatrix} \mathbf{E} \\ \mathbf{V}_{Tc} \end{bmatrix} \qquad (7.7.16)$$

Now, rewrite this in terms of the F submatrices:

$$\mathbf{V}_{Lg} = F_{eg}^{\mathrm{T}} * \mathbf{E} + F_{cg}^{\mathrm{T}} * \mathbf{V}_{Tc} \qquad (7.7.17)$$

This gives us the conductance voltages in terms of the input and the states, so we can now move to getting the state equations.

$$\mathbf{I}_{Tc} = F_{cg} * G * (F_{eg}^{\mathrm{T}} * \mathbf{E} + F_{cg}^{\mathrm{T}} * \mathbf{V}_{Tc}) \qquad (7.7.18)$$

Now, since $\mathbf{I}_{Tc} = C_T * d(\mathbf{V}_{Tc})/dt$, we can solve for the derivative of the state vector, getting the state equations as

$$\frac{d(\mathbf{V}_{Tc})}{dt} = (C_T)^{-1} * F_{cg} * G * ((F_{eg}^{\mathrm{T}} * \mathbf{E}) + (F_{cg}^{\mathrm{T}} * \mathbf{V}_{Tc})) \qquad (7.7.19)$$

Since we have all of the matrices involved in this, we can write out the state equations for this example. Earlier, we defined C_T, G, and we found F_{cg}

and F_{eg}. If we insert the values for these matrices, the state equations are found to be

$$\frac{d(V_{C1})}{dt} = \frac{1}{C_1} * \left(\frac{V_{C1} - E}{R_1} + \frac{V_{C1} - V_{C2}}{R_2} \right) \qquad (7.7.20a)$$

$$\frac{d(V_{C2})}{dt} = \frac{1}{C_2} * \left(\frac{V_{C2} - V_{C1}}{R_2} \right) \qquad (7.7.20b)$$

PROBLEM 7.9

Carry out all of the intermediate steps (matrix manipulations) to get the above state equations.

7.8 GENERATING STATE EQUATIONS FOR NETWORKS WITH VOLTAGE SOURCES, CAPACITORS, RESISTORS, INDUCTORS, AND CURRENT SOURCES

In the preceding section we considered a very specific network that did not have all of the possible kinds of elements that can exist. However, we need to consider a more general formulation of the network equations that can give state equations for a more general network. In this section we will examine a formulation that will handle more general, but not completely general, networks. We will not yet consider, for example, controlled sources. Also, we will continue to restrict consideration to linear networks.

A more general kind of circuit that we will consider in this section is one with the following kinds of components and variables

Tree voltage sources	(E, I_c)
Tree capacitors	(V_{Tc}, I_{Tc})
Tree resistors	(V_{Tr}, I_{Tr})
Tree inductors	(V_{Tl}, I_{Tl})
Link capacitors	(V_{Lc}, I_{Lc})
Link conductances	(V_{Lg}, I_{Lg})
Link inductors	(V_{Ll}, I_{Ll})
Link current sources	(V_j, J)

We assume that there are *no* voltage sources in the links or current sources in the tree. If either of these situations arises, then we are considering a situation that is probably physically impossible. If a voltage source must be in the set of links, then there is a loop of voltage sources, and Kirchhoff's voltage law is most probably violated. If a current source must be placed in the tree, then there is a cut set of current sources, and Kirchhoff's current law is violated. In either case, when such situations are encountered, our program should print an error message and quit.

In reading data in for computer analysis, we will take steps to put the data into the most useful order. That order will be:

Voltage sources
Capacitors
Resistors
Inductors
Current sources

Furthermore, we will order the elements of a given type by element values. Capacitors and resistors will be ordered from largest to smallest, and inductors from smallest to largest. The purpose of this ordering is to ensure that the smallest capacitors will be forced out of the tree as we do the Gaussian reduction to find the tree. In that way, if there are any parasitic capacitances included in the circuit, their voltages will not be taken as states in our final analysis. Similarly, we will want the smallest inductances forced into the tree if there is a cut set of inductors and current sources, so that parasitic inductance currents do not become states later in our analysis.

Before we go further, we should define terms.

$$\mathbf{V}_{\text{link}} = \begin{bmatrix} \mathbf{V}_{Lc} \\ \mathbf{V}_{Lg} \\ \mathbf{V}_{Ll} \\ \mathbf{V}_j \end{bmatrix} \qquad \mathbf{I}_{\text{link}} = \begin{bmatrix} \mathbf{I}_{Lc} \\ \mathbf{I}_{Lg} \\ \mathbf{I}_{Ll} \\ \mathbf{J} \end{bmatrix} \qquad \begin{array}{l} \text{Link capacitors} \\ \text{Link conductances} \\ \text{Link inductors} \\ \text{Link } I \text{ source } V \text{ and } I \end{array} \qquad (7.8.1a,b)$$

$$\mathbf{V}_{\text{tree}} = \begin{bmatrix} \mathbf{E} \\ \mathbf{V}_{Tc} \\ \mathbf{V}_{Tr} \\ \mathbf{V}_{Tl} \end{bmatrix} \qquad \mathbf{I}_{\text{tree}} = \begin{bmatrix} \mathbf{I}_e \\ \mathbf{I}_{Tc} \\ \mathbf{I}_{Tr} \\ \mathbf{I}_{Tl} \end{bmatrix} \qquad \begin{array}{l} \text{Tree } V \text{ source } V \text{ and } I \\ \text{Tree capacitors} \\ \text{Tree resistors} \\ \text{Tree inductors} \end{array} \qquad (7.8.1c,d)$$

These variables are related through Kirchhoff's current law and Kirchhoff's voltage law.

$$[I \vdots F] * \mathbf{I}_{\text{branch}} = 0 \qquad (7.8.2)$$

or

$$\begin{bmatrix} I & 0 & 0 & 0 & : & F_{ec} & F_{eg} & F_{el} & F_{ej} \\ 0 & I & 0 & 0 & : & F_{cc} & F_{cg} & F_{cl} & F_{cj} \\ 0 & 0 & I & 0 & : & F_{rc} & F_{rg} & F_{rl} & F_{rj} \\ 0 & 0 & 0 & I & : & F_{lc} & F_{lg} & F_{ll} & F_{lj} \end{bmatrix} * \begin{bmatrix} \mathbf{I}_{\text{tree}} \\ \mathbf{I}_{\text{link}} \end{bmatrix} = 0 \qquad (7.8.3)$$

A few of the submatrices of F are zero. For example, an inductor gets forced into the tree when the inductor current depends only upon current sources and link inductor currents, because of the existence of an inductor current source cut set. If that tree inductor current depends only upon current in link current sources and upon link inductor currents, then it does not depend upon link capacitor currents and link conductance currents. So, F_{lc} and F_{lg} must be equal to zero. Similarly F_{rc} will be zero because current in tree resistors does not depend upon current in link capacitors.

From Kirchhoff's voltage law, we will also get another set of equations involving F and its submatrices.

$$[-F' \vdots I] * V_{\text{branch}} = 0 \qquad (7.8.4)$$

Together, these two expressions (KCL and KVL) summarize all of the topological information about the network.

We need also to get the information about individual elements into our formulation. We will use a vector matrix formulation for the description of each type of element. Since we will later focus on the states of the system, let us consider the equations that later become state equations.

If there are tree capacitors, then the voltages across the tree capacitors can be taken as state variables. In the vector, V_{tree}, there is a subvector V_{Tc} that contains the voltages across the tree capacitors. Each voltage in V_{Tc} is related to a current in I_{Tc}, a subvector of I_{tree}, the vector of tree branch currents. That relation can be expressed mathematically as follows:

$$\begin{bmatrix} C_a & 0 \cdots & 0 \\ 0 & C_b \cdots & 0 \\ & \vdots & \\ & \vdots & \\ 0 & 0 \cdots & C_n \end{bmatrix} * \begin{bmatrix} dV_a/dt \\ dV_b/dt \\ \vdots \\ \vdots \\ dV_n/dt \end{bmatrix} = \begin{bmatrix} I_a \\ I_b \\ \vdots \\ \vdots \\ I_n \end{bmatrix} \qquad (7.8.5)$$

This set of equations can be written most compactly in vector matrix format as:

$$C_1 * \frac{d(V_{Tc})}{dt} = I_{Tc} \qquad (7.8.6)$$

Similarly, for the link inductor currents, we get state variables, and the key equation is the one relating link inductor current derivatives to link inductor voltages. Written in vector matrix form, this becomes

$$L_1 * \frac{d(I_{Ll})}{dt} = V_{Ll} \qquad (7.8.7)$$

Here, L_1 is a diagonal matrix of link inductor values, and we have

$$I_{Ll} \quad \text{A vector of link inductor currents} \qquad (7.8.8)$$
$$V_{Ll} \quad \text{A vector of link inductor voltages} \qquad (7.8.9)$$

So far we have gotten information about current voltage relations for the states (tree capacitor voltages and link inductor currents) in a form amenable to further manipulation. We need to get such relations for all of the elements. Listed below is a summary of all the element relations we will need:

$C_1 * d(V_{Tc})/dt = I_{Tc}$	Tree capacitors	(7.8.10a)
$L_1 * d(I_{Ll})/dt = V_{Ll}$	Link inductors	(7.8.10b)
$C_2 * d(V_{Lc})/dt = I_{Lc}$	Link capacitors	(7.8.10c)
$L_2 * d(I_{Tl})/dt = V_{Tl}$	Tree inductors	(7.8.10d)
$V_{Tr} = R * I_{Tr}$	Tree resistors	(7.8.10e)
$I_{Lg} = G * V_{Lg}$	Link conductances	(7.8.10f)
$V_{Te} = E$	Tree voltage sources	(7.8.10g)
$I_{Lj} = J$	Link current sources	(7.8.10h)

Here each variable or derivative is premultiplied by a diagonal matrix of some appropriate element values. For example, C_1 is a diagonal matrix of capacitor values for capacitors in the tree, and R is a diagonal matrix of resistance values for resistances in the tree. The vectors $\mathbf{E}$ and $\mathbf{J}$ contain the values of the voltage and current sources. If they are functions, we will need to "tag" these locations to be able to locate the inputs, but for now we assume that $\mathbf{E}$ and $\mathbf{J}$ are vectors of numbers.

The preceding relations integrate all of the information about element values. We need to integrate that, in turn with the topological information contained in Kirchhoff's current and voltage laws:

$$[I \ \vdots \ F] * \mathbf{I}_{\text{branch}} = 0 \qquad (7.8.11a)$$

$$[-F^{\text{T}} \ \vdots \ I] * \mathbf{V}_{\text{branch}} = 0 \qquad (7.8.11b)$$

The state vector for our network will consist of the capacitor voltages for all tree capacitors and the inductor currents for all link inductors. Link capacitor voltages and tree inductor currents are not states. Form the state vector

$$\mathbf{X} = \begin{bmatrix} \mathbf{V}_{Tc} \\ \mathbf{I}_{Ll} \end{bmatrix} \qquad (7.8.12)$$

Then we have two equations for the state derivatives that can be combined into one precursor to the state equation:

$$\begin{bmatrix} C_1 & 0 \\ 0 & L_1 \end{bmatrix} \frac{d}{dt} \begin{bmatrix} \mathbf{V}_{Tc} \\ \mathbf{I}_{Ll} \end{bmatrix} = \begin{bmatrix} \mathbf{I}_{Tc} \\ \mathbf{V}_{Ll} \end{bmatrix} = \begin{bmatrix} -F_{cc}\mathbf{I}_{Lc} & -F_{cg}\mathbf{I}_{Lg} & -F_{cl}\mathbf{I}_{Ll} & -F_{cj}\mathbf{J} \\ F_{ll}^{\text{T}}\mathbf{V}_{Tl} & +F_{rl}^{\text{T}}\mathbf{V}_{Tr} & +F_{cl}^{\text{T}}\mathbf{V}_{Tc} & +F_{ll}^{\text{T}}\mathbf{E} \end{bmatrix} \qquad (7.8.13)$$

From now on use $\mathbf{X}$ for the states and $\mathbf{U}$ for the inputs. We can rewrite this in a way that separates the state, inputs and the other variables.

$$\begin{bmatrix} C_1 & 0 \\ 0 & L_1 \end{bmatrix} \frac{d\mathbf{X}}{dt} = \begin{bmatrix} 0 & -F_{cl} \\ F_{cl}^{\text{T}} & 0 \end{bmatrix} \mathbf{X} + \begin{bmatrix} 0 & -F_{cj} \\ F_{el}^{\text{T}} & 0 \end{bmatrix} \mathbf{U}$$

$$+ \begin{bmatrix} 0 & -F_{cg} \\ F_{rl}^{\text{T}} & 0 \end{bmatrix} \begin{bmatrix} \mathbf{V}_{Tr} \\ \mathbf{I}_{Lg} \end{bmatrix} + \begin{bmatrix} 0 & -F_{cc} \\ F_{ll}^{\text{T}} & 0 \end{bmatrix} \begin{bmatrix} \mathbf{V}_{Tl} \\ \mathbf{I}_{Lc} \end{bmatrix} \qquad (7.8.14)$$

This expression contains variables other than the states $\mathbf{X}$ and the inputs $\mathbf{U}$. We have to eliminate the resistive variables $\mathbf{V}_{Tr}$ and $\mathbf{I}_{Lg}$ and then the excess variables V_l and I_c. We have

$$\begin{bmatrix} \mathbf{V}_{Tr} \\ \mathbf{I}_{Lg} \end{bmatrix} = \begin{bmatrix} R & 0 \\ 0 & G \end{bmatrix} * \begin{bmatrix} \mathbf{I}_{Tr} \\ \mathbf{V}_{Lg} \end{bmatrix}$$

$$= \begin{bmatrix} R & 0 \\ 0 & G \end{bmatrix} * \begin{bmatrix} -F_{rg}\mathbf{I}_{Lg} & -F_{rl}\mathbf{I}_{Ll} & -F_{rj}\mathbf{J} \\ F_{rg}^{\text{T}}\mathbf{V}_{Tr} & +F_{cg}^{\text{T}}\mathbf{V}_{Tc} & +F_{eg}^{\text{T}}\mathbf{E} \end{bmatrix} \qquad (7.8.15)$$

If we collect all of the terms that involve $\mathbf{I}_{Lg}$ and $\mathbf{V}_{Tr}$ (all the while putting terms into the "standard vectors" we have defined) we have:

$$\begin{bmatrix} \mathbf{V}_{Tr} \\ \mathbf{I}_{Lg} \end{bmatrix} = \begin{bmatrix} I & RF_{rg} \\ -GF_{rg}^{\text{T}} & T \end{bmatrix} \begin{bmatrix} R & 0 \\ 0 & G \end{bmatrix} \left\{ \begin{bmatrix} 0 & -F_{rl} \\ F_{cg}^{\text{T}} & 0 \end{bmatrix} \mathbf{X} + \begin{bmatrix} 0 & -F_{rj} \\ F_{eg}^{\text{T}} & 0 \end{bmatrix} \mathbf{U} \right\} \qquad (7.8.16)$$

This gives us the resistive variables in terms of inputs and states. We need to do the same for the excess variables. The defining relation for the excess variables is

$$\begin{bmatrix} \mathbf{V}_{Tl} \\ \mathbf{I}_{Lc} \end{bmatrix} = \begin{bmatrix} L_2 & 0 \\ 0 & C_2 \end{bmatrix} \frac{d}{dt} \begin{bmatrix} \mathbf{I}_{Tl} \\ \mathbf{V}_{Lc} \end{bmatrix} \tag{7.8.17}$$

We have the Kirchhoff current law relations:

$$\mathbf{I}_{Ll} = -F_{ll}\mathbf{I}_{ll} - F_{lj}\mathbf{J} \tag{7.8.18a}$$

and

$$\mathbf{V}_{Lc} = F_{cc}^{\mathrm{T}}\mathbf{V}_{Tc} + F_{cc}^{\mathrm{T}}\mathbf{E} \tag{7.8.18b}$$

So we must have

$$\begin{bmatrix} \mathbf{V}_{Tl} \\ \mathbf{I}_{Lc} \end{bmatrix} = \begin{bmatrix} L_2 & 0 \\ 0 & C_2 \end{bmatrix} \frac{d}{dt} \left\{ \begin{bmatrix} 0 & -F_{ll} \\ F_{cc}^{\mathrm{T}} & 0 \end{bmatrix} \begin{bmatrix} \mathbf{V}_{Tc} \\ \mathbf{I}_{Ll} \end{bmatrix} + \begin{bmatrix} 0 & -F_{lj} \\ F_{ec}^{\mathrm{T}} & 0 \end{bmatrix} \begin{bmatrix} \mathbf{E} \\ \mathbf{J} \end{bmatrix} \right\} \tag{7.8.19}$$

Finally, we can take these expressions for resistive variables and excess variables and put them back into the precursor of the state equations, obtaining:

$$\begin{bmatrix} C_1 & 0 \\ 0 & L_1 \end{bmatrix} \frac{d\mathbf{X}}{dt} = \begin{bmatrix} 0 & -F_{cl} \\ F_{cl}^{\mathrm{T}} & 0 \end{bmatrix} \mathbf{X} + \begin{bmatrix} 0 & -F_{cj} \\ F_{el}^{\mathrm{T}} & 0 \end{bmatrix} \mathbf{U}$$

$$+ \begin{bmatrix} 0 & -F_{cg} \\ F_{rl}^{\mathrm{T}} & 0 \end{bmatrix} \begin{bmatrix} I & RF_{rg} \\ -GF_{rg}^{\mathrm{T}} & I \end{bmatrix}^{-1} \left\{ \begin{bmatrix} 0 & -RF_{rj} \\ GF_{eg}^{\mathrm{T}} & 0 \end{bmatrix} \mathbf{U} \right.$$

$$+ \begin{bmatrix} 0 & -RF_{rl} \\ GF_{cg}^{\mathrm{T}} & 0 \end{bmatrix} \mathbf{X} \right\} + \begin{bmatrix} 0 & -F_{cc} \\ F_{ll}^{\mathrm{T}} & 0 \end{bmatrix} \begin{bmatrix} L_2 & 0 \\ 0 & C_2 \end{bmatrix} \begin{bmatrix} 0 & -F_{ll} \\ F_{cc}^{\mathrm{T}} & 0 \end{bmatrix} \frac{d\mathbf{X}}{dt}$$

$$+ \begin{bmatrix} 0 & -F_{lj} \\ F_{ec}^{\mathrm{T}} & 0 \end{bmatrix} \frac{d\mathbf{U}}{dt} \right\} \tag{7.8.20}$$

Examining these equations, we find that the inclusion of excess elements, either link capacitors or tree inductors, will add a different term to the right-hand side of the state equation. That term involves derivatives of the inputs. Including that potentially troublesome term and solving for the state derivatives we get

$$\dot{\mathbf{X}} = \begin{bmatrix} C_1 & F_{cc}C_2F_{cc}^{\mathrm{T}} \\ F_{ll}^{\mathrm{T}}L_2F_{ll} & L_1 \end{bmatrix}^{-1} \left(\begin{bmatrix} 0 & -F_{cl} \\ F_{cl}^{\mathrm{T}} & 0 \end{bmatrix} \right.$$

$$+ \begin{bmatrix} 0 & -F_{cg} \\ F_{rl}^{\mathrm{T}} & 0 \end{bmatrix} \begin{bmatrix} I & RF_{rg} \\ -GF_{rg}^{\mathrm{T}} & I \end{bmatrix}^{-1} \begin{bmatrix} 0 & -RF_{rl} \\ GF_{cg}^{\mathrm{T}} & 0 \end{bmatrix} \right) \mathbf{X}$$

$$+ \begin{bmatrix} C_1 & F_{cc}C_2F_{cc}^{\mathrm{T}} \\ F_{ll}^{\mathrm{T}}L_2F_{ll} & L_1 \end{bmatrix}^{-1} \left(\begin{bmatrix} 0 & -F_{cj} \\ F_{el}^{\mathrm{T}} & 0 \end{bmatrix} \right.$$

$$+ \begin{bmatrix} 0 & -F_{cg} \\ F_{rl}^{\mathrm{T}} & 0 \end{bmatrix} \begin{bmatrix} I & RF_{rg} \\ -GF_{rg}^{\mathrm{T}} & I \end{bmatrix}^{-1} \begin{bmatrix} 0 & -RF_{rj} \\ GF_{eg}^{\mathrm{T}} & 0 \end{bmatrix} \right) \mathbf{U}$$

$$+ \begin{bmatrix} C_1 & F_{cc}C_2F_{cc}^{\mathrm{T}} \\ F_{ll}^{\mathrm{T}}L_2F_{ll} & L_1 \end{bmatrix}^{-1} \begin{bmatrix} 0 & -F_{cc}C_2F_{ec}^{\mathrm{T}} \\ -F_{ll}^{\mathrm{T}}L_2F_{lj} & 0 \end{bmatrix} \dot{\mathbf{U}} \tag{7.8.21}$$

The form of the state equation we get is

$$\frac{d\mathbf{X}}{dt} = A * \mathbf{X} + B * \mathbf{U} + B_1 * \frac{d\mathbf{U}}{dt} \tag{7.8.22}$$

The inclusion of the $d\mathbf{U}/dt$ term, derivatives of the input(s), comes about only when we have excess elements. The presence of an input derivative term complicates the state equations in a way that is not easily gotten around when we want to simulate. Using general integration algorithms (such as Runge-Kutta or Gear algorithms) becomes much more difficult with the input derivative term, since those general algorithms can only accommodate that term by doing a numerical differentiation (assuming that no expression is available for evaluating the derivative—the usual case).

7.9 THE OUTPUT EQUATIONS

Although we have the equations for the derivatives of the states of the system, we also need to derive expressions for the outputs in terms of the states. In this section we will give a derivation of the output expressions.

Let us assume that we have a list of outputs composed only of voltages across particular elements and currents through other elements. (In other words, there is no output that is just a voltage difference between two nodes not directly connected by an element). Each output is either a voltage across an element or a current through an element. Then the output vector, **Y**, can be written as

$$\mathbf{Y} = \gamma \begin{bmatrix} \mathbf{V}_T \\ \cdots \\ \mathbf{V}_L \end{bmatrix} + \delta \begin{bmatrix} \mathbf{I}_T \\ \cdots \\ \mathbf{I}_L \end{bmatrix} = [\gamma_T \vdots \gamma_L] \begin{bmatrix} \mathbf{V}_T \\ \cdots \\ \mathbf{V}_L \end{bmatrix} + [\delta_T \vdots \delta_L] \begin{bmatrix} \mathbf{I}_T \\ \cdots \\ \mathbf{I}_L \end{bmatrix} \tag{7.9.1}$$

Here we have partitioned the variables into tree and link variables. Each output is either voltage across a component or current through a component, so all entries in the matrices are either $+1$ or -1. The output relations can be modified to take topological information into account.

$$\begin{aligned} \mathbf{Y} &= [\gamma_T \mathbf{V}_T + \gamma_L \mathbf{V}_L] + [\delta_T \mathbf{I}_T + \delta_L \mathbf{I}_L] \\ &= [\gamma_T \mathbf{V}_T + \delta_L F^T \mathbf{V}_T] + [-\delta_T F \mathbf{I}_L + \delta_L \mathbf{I}_L] \\ &= [\gamma_T + \gamma_L F^T] \mathbf{V}_T + [-\delta_T F + \delta_L] \mathbf{I}_L \end{aligned} \tag{7.9.2}$$

Define two new quantities and subdivide into appropriate submatrices to get an expression for the output.

$$\zeta = [\gamma_T + \gamma_L F^T] \qquad \eta = [-\delta_T F + \delta_L] \tag{7.9.3a,b}$$

$$\mathbf{Y} = [\zeta_E \mathbf{E} + \zeta_c \mathbf{V}_{Tc} + \zeta_R \mathbf{V}_{Tr} + \zeta_L \mathbf{V}_{Tl}] + [\eta_c \mathbf{I}_{Lc} + \eta_G \mathbf{I}_{Lg} + \eta_L \mathbf{I}_{Ll} + \eta_J \mathbf{J}]$$

$$= [\zeta_E \eta_J] \mathbf{U} + [\zeta_C \vdots \eta_L] \mathbf{X} + [\zeta_R \eta_G] \begin{bmatrix} \mathbf{V}_{Tr} \\ \mathbf{I}_{Lg} \end{bmatrix} + [\zeta_L \vdots \eta_c] \begin{bmatrix} \mathbf{V}_{Tl} \\ \mathbf{I}_{Lc} \end{bmatrix} \tag{7.9.4}$$

Now, we can use the expressions for the resistive variables as developed in the last section. We find the following.

$$
\mathbf{Y} = [\zeta_e \vdots \eta_j]\mathbf{U} + [\zeta_c \vdots \eta_l]\mathbf{X} + [\zeta_r \vdots \eta_g]\begin{bmatrix} I & RF_{rg} \\ -GF_{rg}^{\mathrm{T}} & I \end{bmatrix}^{-1}\left(\begin{bmatrix} 0 & -RF_{rj} \\ GF_{eg}^{\mathrm{T}} & 0 \end{bmatrix}\mathbf{U}\right.
$$

$$
+ \begin{bmatrix} 0 & -RF_{rl} \\ GF_{cg}^{\mathrm{T}} & 0 \end{bmatrix}\mathbf{X}\right) + [\zeta_l \vdots \eta_c]\begin{bmatrix} L_2 & 0 \\ 0 & C_2 \end{bmatrix}\frac{d}{dt}\left(\begin{bmatrix} 0 & -F_{ll} \\ F_{cc}^{\mathrm{T}} & 0 \end{bmatrix}\mathbf{X}\right.
$$

$$
+ \begin{bmatrix} 0 & F_{lj} \\ F_{ec}^{\mathrm{T}} & 0 \end{bmatrix}\mathbf{U}\right)
\tag{7.9.5}
$$

$$
\mathbf{Y} = [\zeta_e \vdots \eta_j]\mathbf{U} + [\zeta_c \vdots \eta_l]\mathbf{X} + [\zeta_r \vdots \eta_g]\begin{bmatrix} I & RF_{rg} \\ -GF_{rg}^{\mathrm{T}} & I \end{bmatrix}^{-1}\begin{bmatrix} 0 & -RF_{rj} \\ GF_{eg}^{\mathrm{T}} & 0 \end{bmatrix}\mathbf{U}
$$

$$
+ [\zeta_l \vdots \eta_c]\begin{bmatrix} L_2 & 0 \\ 0 & C_2 \end{bmatrix}\begin{bmatrix} 0 & -F_{ll} \\ F_{cc}^{\mathrm{T}} & 0 \end{bmatrix}B\mathbf{U}
$$

$$
+ [\zeta_r \vdots \eta_g]\begin{bmatrix} I & RF_{rg} \\ -GF_{rg}^{\mathrm{T}} & I \end{bmatrix}^{-1}\begin{bmatrix} 0 & -RF_{rl} \\ GF_{cg}^{\mathrm{T}} & 0 \end{bmatrix}\mathbf{X}
$$

$$
+ [\zeta_l \vdots \eta_c]\begin{bmatrix} L_2 & 0 \\ 0 & C_2 \end{bmatrix}\begin{bmatrix} 0 & -F_{ll} \\ F_{cc}^{\mathrm{T}} & 0 \end{bmatrix}A\mathbf{X}
$$

$$
+ [\zeta_l \vdots \eta_c]\begin{bmatrix} L_2 & 0 \\ 0 & C_2 \end{bmatrix}\left(\begin{bmatrix} 0 & -F_{ll} \\ F_{cc}^{\mathrm{T}} & 0 \end{bmatrix}B_1\frac{d^2\mathbf{U}}{dt^2} + \begin{bmatrix} 0 & F_{lj} \\ F_{ec}^{\mathrm{T}} & 0 \end{bmatrix}\frac{d\mathbf{U}}{dt}\right)
\tag{7.9.6}
$$

The net result of these manipulations is a set of output equations of the form:

$$
\mathbf{Y} = C*\mathbf{X} + D*\mathbf{U} + D_1*(d\mathbf{U}/dt) + D_2*(d^2\mathbf{U}/dt^2)
\tag{7.9.7}
$$

We have

$$
C = [\zeta_c \vdots \eta_l] + [\zeta_r \vdots \eta_g]\begin{bmatrix} I & RF_{rg} \\ -GF_{rg}^{\mathrm{T}} & I \end{bmatrix}^{-1}\begin{bmatrix} 0 & -RF_{rl} \\ GF_{cg}^{\mathrm{T}} & 0 \end{bmatrix}
$$

$$
+ [\zeta_l \vdots \eta_c]\begin{bmatrix} L_2 & 0 \\ 0 & C_2 \end{bmatrix}\begin{bmatrix} 0 & -F_{ll} \\ F_{cc}^{\mathrm{T}} & 0 \end{bmatrix}A
\tag{7.9.8}
$$

$$
D = [\zeta_e \vdots \eta_j] + [\zeta_r \vdots \eta_g]\begin{bmatrix} I & RF_{rg} \\ -GF_{rg}^{\mathrm{T}} & I \end{bmatrix}^{-1}\begin{bmatrix} 0 & -RF_{rj} \\ GF_{eg}^{\mathrm{T}} & 0 \end{bmatrix}
$$

$$
+ [\zeta_l \vdots \eta_c]\begin{bmatrix} L_2 & 0 \\ 0 & C_2 \end{bmatrix}\begin{bmatrix} 0 & -F_{ll} \\ F_{cc}^{\mathrm{T}} & 0 \end{bmatrix}B
\tag{7.9.9}
$$

$$
D_1 = [\zeta_l \vdots \eta_c]\begin{bmatrix} L_2 & 0 \\ 0 & C_2 \end{bmatrix}\begin{bmatrix} 0 & -F_{lj} \\ F_{ec}^{\mathrm{T}} & 0 \end{bmatrix}
\tag{7.9.10}
$$

$$
D_2 = [\zeta_l \vdots \eta_c]\begin{bmatrix} L_2 & 0 \\ 0 & C_2 \end{bmatrix}\begin{bmatrix} 0 & -F_{ll} \\ F_{cc}^{\mathrm{T}} & 0 \end{bmatrix}B_1
\tag{7.9.11}
$$

With these expressions we have the complete set of state equations that describe our network.

7.10 CIRCUITS WITH CONTROLLED SOURCES

Controlled sources occur frequently in circuits, usually in the more interesting and important ones. However, they present unusual problems when trying to generate state equations. In general, it is not possible to be sure that a set of state equations can even be generated if there are controlled sources in a network, since controlled sources can lead to degeneracies and complications in the analysis of any particular circuit. A few examples will show some of the peculiarities that can occur.

EXAMPLE 7.1

The circuit below has three capacitors but has only one state. Two of the capacitors are in capacitor-voltage source loops, and the voltages across them are not state variables. One of the voltage sources (V_3) is controlled by the current in capacitor C_2. The current source is controlled by the current in capacitor C_4.

We can calculate the derivative of the state (the output voltage in this case) in terms of other variables, carrying our computations back through the circuit to the input voltage.

$$C_6 \frac{dV_6}{dt} = -i_5 = +K_5 i_4 \qquad \text{(CCCS)}$$

$$= +K_5 C_4 \frac{dV_4}{dt} = +K_5 C_4 \frac{dV_3}{dt}$$

$$= +K_5 C_4 \frac{d(K_3 i_2)}{dt} \qquad \text{(CCVS)}$$

$$= +K_5 C_4 K_3 \frac{d}{dt}\left(C_2 \frac{dV_s}{dt}\right)$$

$$= +K_5 C_4 K_3 C_2 \frac{d^2 V_s}{dt^2}$$

The output voltage, V_6, is a state here, but the state derivative involves the second derivative of the input voltage. Clearly this is not of a form that we have encountered before, or that any development to this point would prepare us to handle.

PROBLEM 7.10

Starting with the circuit in Example 7.1, devise a circuit in which the single state derivative depends upon the third derivative of an input voltage.

PROBLEM 7.11

SPICE has an interesting reaction to the circuit in Example 7.1. Write a SPICE program for that network, and execute the program. The job should abort. Determine what changes you have to make in the circuit to get it to run, and then determine whether you have removed the degeneracy in the circuit that has caused the problems above. (Note that you will have to include some zero-voltage voltage sources in order to make the two current-controlled sources work!)

The situation becomes even more intractable if we permit feedback loops within a circuit.

EXAMPLE 7.2

The circuit below has a feedback loop, but is based on the circuit in the last example.

$$C_4 \frac{dV_4}{dt} = -i_3 = K_3 i_2 \qquad \text{(CCCS)}$$

$$= K_3 C_2 \frac{dV_2}{dt} = K_3 C_2 \frac{dV_1}{dt}$$

$$= K_3 C_2 K_1 \frac{dV_4}{dt} \qquad \text{(Only possible if } C_4 = K_3 C_2 K_1)$$

As the derivation shows, only one choice of component values produces equality, even though we can clearly pick any component values we want!

The problem that has developed here comes about when excess elements (capacitors forced out of the proper tree, or inductors forced into the proper tree) combine with controlled sources. Without controlled sources we have found that the state equations might contain derivatives of independent sources. With controlled sources the state equations can contain source derivatives of any order.

Besides the peculiar behavior that introduces higher-order derivatives into the state equations, the presence of controlled sources can also add a great deal of confusion when we try to determine just how many states there are in a circuit. The next example amply illustrates that point.

EXAMPLE 7.3

In the circuit shown, the voltage-controlled voltage source is used to cancel out the inductor voltage exactly. Then the inductor current is just V/R, and the circuit apparently has no real states.

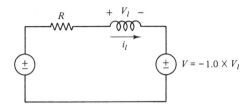

Clearly the situation here is out of hand. It will be extraordinarily difficult to complete any sort of general derivation of state equations in this situation. All we can do is consider some special cases, and defer. However, there is an approach that can illustrate how state equations can be obtained for those networks that are "amenable."

Consider a network that contains controlled sources. Each controlled source can, for the moment, be considered to be an independent source input to the network. Each controlled source is controlled by a network variable (current or voltage) elsewhere in the network. Each controlled variable can, for the moment, be considered to be an output. Each controlled source, in effect, is a network input that is proportional to some network output, in effect a feedback situation. We can analyze the circuit in feedback terms to determine how the circuit behaves. A simple example will help to clarify what we mean here.

EXAMPLE 7.4

The circuit shown has two voltage sources, one of which is controlled by the voltage across the first capacitor. We can write state equations in which both voltage sources appear as inputs, and in which both capacitor voltages (including the first capacitor voltage, which controls the second voltage source) appear as outputs. These equations are as follows:

$$\dot{\mathbf{X}} = \begin{bmatrix} -1/\tau_1 & 0 \\ 0 & -1/\tau_2 \end{bmatrix}\mathbf{X} + \begin{bmatrix} 1/\tau_1 \\ 0 \end{bmatrix}V_1 + \begin{bmatrix} 0 \\ 1/\tau_2 \end{bmatrix}V_2$$

$$V_2 = [g \quad 0]\mathbf{X} \qquad \dot{\mathbf{X}} = \begin{bmatrix} -1/\tau_1 & 0 \\ g/\tau_2 & -1/\tau_2 \end{bmatrix}\mathbf{X} + \begin{bmatrix} 1/\tau_1 \\ 0 \end{bmatrix}V_1$$

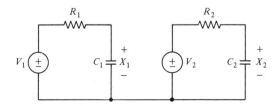

We may write a more general set of state equations to include controlled voltage sources as inputs.

$$\dot{\mathbf{X}} = A\mathbf{X} + [B \; \vdots \; B_c]\begin{bmatrix}\mathbf{U}_r \\ \mathbf{U}_c\end{bmatrix} \tag{7.10.1a}$$

$$\mathbf{Y} = \begin{bmatrix}\mathbf{Y}_r \\ \mathbf{Y}_c\end{bmatrix} = [C \; \vdots \; C_c]\mathbf{X} + \begin{bmatrix}D_{rr} & D_{rc} \\ D_{cr} & D_{cc}\end{bmatrix}\begin{bmatrix}\mathbf{U}_r \\ \mathbf{U}_c\end{bmatrix} \tag{7.10.1b}$$

$$\mathbf{U}_c = G\mathbf{Y}_c \tag{7.10.1c}$$

In these equations the controlled sources ($\mathbf{U}_c$) and the outputs (pseudo-outputs?) that control the controlled sources are separated out from the "real" inputs and outputs. Note that the state equations above assume that we have no excess elements (no capacitor–voltage-source loops, and no inductor–current-source cut sets) since only the input (and not the derivative of the input) appears in the state equations.

We can manipulate all of these equations to get a single state equation that accounts for the presence of the controlled sources.

$$\dot{\mathbf{X}} = A\mathbf{X} + B_r\mathbf{U}_r + B_c\mathbf{U}_c \tag{7.10.2}$$

To solve for $\mathbf{U}_c$:

$$\mathbf{U}_c = G\mathbf{Y}_c = GC_c\mathbf{X} + GD_{cr}\mathbf{U}_r + GD_{cc}\mathbf{U}_c$$
$$[I - GD_{cc}]\mathbf{U}_c = GC_c\mathbf{X} + GD_{cr}\mathbf{U}_r$$
$$\mathbf{U}_c = [I - GD_{cc}]^{-1}[GC_c\mathbf{X} + GD_{cr}\mathbf{U}_r] \tag{7.10.3}$$

So

$$\dot{\mathbf{X}} = A\mathbf{X} + B_r\mathbf{U}_r + B_c\mathbf{U}_c$$
$$= A\mathbf{X} + B_r\mathbf{U}_r + B_c[I - GD_{cc}]^{-1}[GC_c\mathbf{X} + GD_{cr}\mathbf{U}_r]$$
$$= [A + B_c[I - GD_{cc}]^{-1}GC_c]\mathbf{X} + [B_r + B_c[I - GD_{cc}]^{-1}GD_{cr}]\mathbf{U}_r \tag{7.10.4}$$

When we attempt to solve for the outputs that control the controlled sources, we find we have to invert a matrix, $[I - GD_{cc}]$, and we have no a priori assurance that the inverse exists. It will not exist when zero memory loops exist that involve the controlled sources. This is the first possible way for this analysis to fail. With excess elements, other failure modes become possible. Note that the controlled sources do not produce any input derivative terms in this formulation for no excess elements.

If we have excess elements, the situation can change. With excess elements, the state equations take the form

$$\dot{\mathbf{X}} = A\mathbf{X} + B\mathbf{U} + B_1\dot{\mathbf{U}} \tag{7.10.5a}$$
$$\mathbf{Y} = C\mathbf{X} + D\mathbf{U} + D_1\dot{\mathbf{U}} + D_2\ddot{\mathbf{U}} \tag{7.10.5b}$$

Partitioning these equations to account for real and controlled sources, we get

$$\mathbf{Y} = \begin{bmatrix}\mathbf{Y}_r \\ \mathbf{Y}_c\end{bmatrix} = [C_r \quad C_c]\mathbf{X} + \begin{bmatrix}D_{rr} & D_{rc} \\ D_{cr} & D_{cc}\end{bmatrix}\begin{bmatrix}\mathbf{U}_r \\ \mathbf{U}_c\end{bmatrix} + \begin{bmatrix}D_{1rr} & D_{1rc} \\ D_{1cr} & D_{1cc}\end{bmatrix}\begin{bmatrix}\dot{\mathbf{U}}_r \\ \dot{\mathbf{U}}_c\end{bmatrix}$$
$$+ \begin{bmatrix}D_{2rr} & D_{2rc} \\ D_{2cr} & D_{2cc}\end{bmatrix}\begin{bmatrix}\ddot{\mathbf{U}}_r \\ \ddot{\mathbf{U}}_c\end{bmatrix} \tag{7.10.6}$$

Substituting back and solving for the "pseudo-outputs" due to the controlled sources, we get

$$\mathbf{U}_c = G\mathbf{Y}_c = C_c\mathbf{X} + D_{cr}\mathbf{U}_r + D_{1cr}\dot{\mathbf{U}}_r + D_{2cr}\ddot{\mathbf{U}}_r + D_{cc}G\mathbf{Y}_c$$
$$+ D_{1cc}G\dot{\mathbf{Y}}_c + D_{2cc}\ddot{\mathbf{Y}}_c \qquad (7.10.7)$$

At this point we may have reached a dead end for our purposes. If the derivative term, $d\mathbf{Y}_c/dt$, is actually present, then some of the excess elements (which were assumed not to comprise states) are actually contributing to this subsidiary differential equation. If that is the case, we have the situation discussed in Example 7.1 in which the state equation begins to have higher-order derivatives of the input. We leave that for other texts, but we do note that if the derivative term, $d\mathbf{Y}_c/dt$, has any nonzero elements, that is one way of defining a network beyond what our methods will handle. Otherwise, we might be able to proceed.

There is an important class of networks that can cause problems. A circuit with mutual inductance is sometimes most easily handled through a conversion to an equivalent circuit with a cut set of inductors. Skilling [5, pp. 265–267] shows how mutual inductances can be converted to three inductors forming a cut set and an ideal transformer (which is really a voltage-controlled voltage source). Hostetter [6, pp. 210–212] presents a controlled current source equivalent for mutual inductances, but the controlling variable is a derivative of a current. Some browsing through any number of basic circuit texts, with the preceding development in mind will provide fairly convincing arguments that mutual inductors can be difficult to handle. Since mutual inductors are still a common circuit element, some caution is advisable when they are encountered and a circuit analysis package is used for analyzing such a circuit.

7.11 A SUMMARY AND A PREVIEW

In this chapter we have derived a state equation description of a network starting from a topological description of the network. That set of state equations is in a standard form that can be used to generate other useful descriptions of the network.

In the next chapters we will look at ways of going from the state equation description of the network to a transfer function or frequency response description. First we will spend some time discussing how to calculate transient response of the network—that is, the network's response from initial conditions or the response caused by some specified input.

Our derivation of the state equation description assumed that the network was linear. It is possible, though difficult, to make the transition from topological description to state equations for nonlinear networks. However, if a network is linear (as is often a reasonable assumption), taking advantage of the network's linearity can lead to considerable simplification and speedup in the computations involved. The same generalizations hold true in computation of the transient response. In the next chapter we will look at methods appropriate for computation of linear system response.

Our derivation is not completely general, since we have not covered thoroughly all of the special cases with dependent sources. Those particular cases can cause trouble in any circuit analysis program since it is possible to construct circuits that have arbitrarily high degrees of input derivatives (for example) appearing in the state equations.

The material in this chapter is not material that you will use very often; however, the understanding of types of problem circuits and how they arise is what is useful. You may never write a circuit analysis program. You will almost certainly use several circuit analysis programs, and you will need to be aware of the pitfalls.

REFERENCES

1. C. A. Desoer and E. S. Kuh, *Basic Circuit Theory,* McGraw-Hill, New York, 1969. (Chapter 11 of this text gives a clear and readable exposition of the concepts of topological analysis applied to electric circuits. Well worth reading through!)
2. N. Balabanian and W. R. LePage, *Electrical Science, Book 1—Resistive and Diode Networks,* McGraw-Hill, New York, 1970. (Chapter 7 of this very readable programmed text gives an introduction to topological concepts including graphs, cut sets, and incidence matrices.)
3. D. A. Calahan, *Computer-Aided Network Design,* rev. ed., McGraw-Hill, New York, 1972.
4. L. O. Chua and P-M Lin, *Computer Aided Analysis of Electronic Circuits: Algorithms and Computational Techniques,* Prentice-Hall, Englewood Cliffs, NJ, 1975. (Chapter 8 of this book gives a complete exposition of the derivation of the state equations for active networks. Good, but terse.)
5. H. H. Skilling, *Electric Networks,* Wiley, New York, 1974.
6. G. H. Hostetter, *Engineering Network Analysis,* Harper & Row, New York, 1984.

MINIPROJECTS

One of the most useful items of information you can have on any prepackaged program is the limitations of its usefulness. In this chapter we have discussed the areas in which problems can arise. In these miniprojects we will examine several of those areas. We assume that some sort of package is available at your location and that you will use that package in solving the problems presented in these miniprojects.

In each project, test what happens when that particular class of circuit is input to your package. Provide a theoretical justification of the behavior of the package.

For each class of circuit, investigate whether or not the class of circuit poses a problem for your particular circuit package. If the class of circuit does indeed pose a problem, then place yourself in the position of the writer of the instruction manual for the package and prepare a write up that (a) describes the type of circuit that causes the problem, and (b) explains to the user either what the theoretical problem is and/or how to change the circuit to eliminate the problem.

1. Our development assumed that the circuit being analyzed had a tree. Disjoint circuits have no trees. Below are two circuits. The upper circuit has a ground connection in both parts; the lower circuit does not.

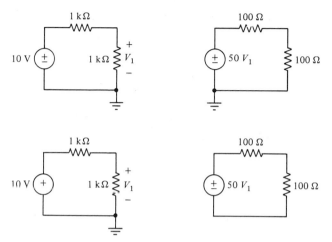

2. The circuit below can become troublesome for one particular value of transconductance. Find the value and proceed as outlined above.

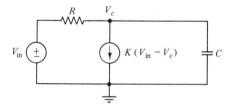

3. Isolation transformers are frequently used to provide isolation from dangerous voltage levels. A simple application of an isolation transformer is shown below.

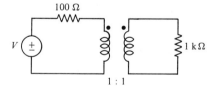

Chapter 8

Transient Response Using Matrix Exponentials

8.1 INTRODUCTION

When it is possible to take advantage of a special knowledge of a system, it may be possible to obtain some sort of advantage. In Chapters 3 and 4 we considered some general methods of computing the transient response of a circuit or system. General algorithms, like the Euler or Runge-Kutta methods, assume that very little is known about the system. However, if we have some detailed knowledge of the system and can use that knowledge, then we may be able to improve our computations by using less CPU time, doing a more accurate computation, or doing a simpler computation. In the preceding chapter we considered computer generation of state equations for linear, time-invariant electrical networks. That is a fairly restricted type of system, and it should prove possible to take advantage of the knowledge of the form of the circuits we have considered.

The general kind of system we deal with is described by the vector matrix differential equation

$$\dot{\mathbf{X}} = A * \mathbf{X} + B * \mathbf{U} \tag{8.1.1}$$

This equation describes a general linear, time-invariant system (one without time lags). We assume we have N states, in general. However, to introduce the matrix exponential, consider a system with just a single state, x. This system will be described by the scalar state equation:

$$\dot{x} = ax + bu \tag{8.1.2}$$

Furthermore, assume that the input, u, is always zero. In this situation, the only way we can have anything happening in the system is if we have some sort of nonzero initial condition. Call the initial condition, $x(0)$. Then the solution of the scalar undriven state equation is given by

$$x(t) = e^{at}x(0) \tag{8.1.3}$$

What do we mean when we write this solution? In particular, what do we mean by e^{at}? We all have some familiarity with this exponential function, but what exactly do we mean when we use that function? If we momentarily ignore all the lore about e being the base for natural logarithms, we realize that the exponential function is defined as the sum of an infinite series:

$$e^{at} = 1 + at + \frac{(at)^2}{2} + \frac{(at)^3}{3!} + \cdots \tag{8.1.4}$$

If we use this series in the solution wherever e^{at} occurs, we can gain a little more of the insight we need. Substituting the series in our solution, we find

$$x(t) = \left[1 + at + \frac{(at)^2}{?} + \frac{(at)^3}{3!} + \cdots \right]x(0) \tag{8.1.5}$$

Then, if we differentiate this solution, we get

$$\dot{x}(t) = \left[0 + a + a(at) + \frac{a(at)^2}{2} + \cdots \right]x(0)$$
$$= ax(t) \tag{8.1.6}$$

In other words, by using the series we can see directly that the series solution satisfies the differential equation with zero input.

We can generalize this series solution to the case of a vector state. In that case, we can hypothesize that the solution to the undriven (zero-input) state equation is given by

$$\mathbf{X}(t) = \left[I + At + \frac{(At)^2}{2} + \frac{(At)^3}{3!} + \cdots \right] * \mathbf{X}(0) \tag{8.1.7}$$

In the case of a vector of states, we must replace "a" with the matrix A, and "1" by the identity matrix, I. Then, this looks similar to the series solution we had previously. Then, if we compute the derivative of the state vector, we have

$$\dot{\mathbf{X}}(t) = \left[0 + A + At + \frac{(At)^2}{2} + \cdots \right] * \mathbf{X}(0) \tag{8.1.8}$$
$$= A * \mathbf{X}(t) \tag{8.1.9}$$

Again, the series solution satisfies the differential equation and we take it as the unique solution.

By analogy with the ordinary exponential, e^{at}, we define the series above as e^{At}. Normally, this is referred to as the *matrix exponential*. We have to be more than a little careful with that term. We can't really mean that we take e, the natural logarithm base to a matrix power, since that concept doesn't seem to make much sense. For now, let us just take the term "matrix exponential" to mean this series, and try to avoid associating that term in our mind with the idea of raising a number to a power (and a matrix power at that).

The concept of a matrix exponential is useful any time we need to analyze a linear time invariant system described by a set of state equations [1, 2]. It arises in systems with inputs, and it arises in a wide variety of types of systems. See Patten [3] for an example of matrix exponentials in biological systems, for example.

The matrix exponential also appears in analytical solutions using transform techniques. Assume that we have an undriven (autonomous) linear system:

$$\dot{\mathbf{X}} = A * \mathbf{X} \tag{8.1.10}$$

Then, taking Laplace transforms and accounting for an initial state, x(0), we find

$$s\mathbf{X}(s) - \mathbf{X}(0) = A * \mathbf{X}(s) \tag{8.1.11}$$

or
$$(sI - A) * \mathbf{X}(s) = \mathbf{X}(0) \tag{8.1.12}$$

$$\mathbf{X}(s) = (sI - A)^{-1} * \mathbf{X}(0) \tag{8.1.13}$$

This has to correspond to the time domain expression

$$\mathbf{X}(t) = e^{At} * \mathbf{X}(0) \tag{8.1.14}$$

Equations (8.1.13) and (8.1.14) are alternative descriptions of the same situation. Comparing these two expressions, we note that the matrix exponential, e^{At} must be the inverse Laplace transform of $(sI - A)^{-1}$. Our next example illustrates this relationship between e^{At} and $(sI - A)^{-1}$.

The relationship that exists between the matrix exponential and its Laplace transform means that we can take advantage of what we know about one to learn about the other. For example, every element of the inverse of $(sI - A)$ has a denominator that is the determinant of $(sI - A)$. In turn, $|sI - A|$ is a polynomial in s. That polynomial is the *characteristic polynomial* of the matrix A, and is of degree N when A is N by N. Roots of the characteristic polynomial (eigenvalues of the matrix, A) are the poles of the system. These relationships can be made clearer by considering an example.

EXAMPLE 8.1
A block diagram of a system with two time constants is shown.

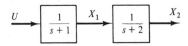

The state equation for this system is

$$\dot{\mathbf{X}} = \begin{bmatrix} -1 & 0 \\ 1 & -2 \end{bmatrix} * \mathbf{X} + \begin{bmatrix} 1 \\ 0 \end{bmatrix} * \mathbf{U}$$

For this system, we have

$$A = \begin{bmatrix} -1 & 0 \\ 1 & -2 \end{bmatrix}$$

So $\qquad (sI - A)^{-1} = \begin{bmatrix} s+1 & 0 \\ -1 & s+2 \end{bmatrix}^{-1}$

$$= \begin{bmatrix} \dfrac{s+2}{(s+1)(s+2)} & \dfrac{1}{(s+1)(s+2)} \\ 0 & \dfrac{s+1}{(s+1)(s+2)} \end{bmatrix}^{T}$$

$$= \begin{bmatrix} \dfrac{1}{s+1} & 0 \\ \dfrac{1}{(s+1)(s+2)} & \dfrac{1}{s+2} \end{bmatrix}$$

If we take the inverse Laplace transform of $(sI - A)^{-1}$, we have

$$e^{At} = \begin{bmatrix} e^{-t} & 0 \\ e^{-t} - e^{-2t} & e^{-2t} \end{bmatrix}$$

EXAMPLE 8.2

The same transfer function as in Example 8.1 can be realized with another block diagram, as shown.

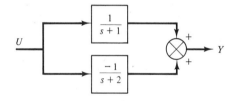

The state equation for this system is

$$\dot{X} = \begin{bmatrix} -1 & 0 \\ 0 & -2 \end{bmatrix} * X + \begin{bmatrix} 1 \\ -1 \end{bmatrix} * U$$

For this system we have

$$A = \begin{bmatrix} -1 & 0 \\ 0 & -2 \end{bmatrix}$$

Thus $\qquad (sI - A)^{-1} = \begin{bmatrix} s+1 & 0 \\ 0 & s+2 \end{bmatrix}^{-1} = \begin{bmatrix} \dfrac{1}{s+1} & 0 \\ 0 & \dfrac{1}{s+2} \end{bmatrix}$

If we take the inverse Laplace transform of $(sI - A)^{-1}$, we obtain

$$e^{At} = \begin{bmatrix} e^{-t} & 0 \\ 0 & e^{-2t} \end{bmatrix}$$

PROBLEM 8.1

Find the matrix exponential for the system described by the block diagram shown.

U → $\dfrac{20}{(s + 2)(s + 10)}$ → Y

From these examples and Problem 8.1 it should be apparent that the same exponential functions turn up in the matrix exponential, regardless of the particular representation chosen for the system. All that matters is the location of the poles of the system. The poles determine the particular exponential functions that appear, although the representation chosen will affect how those exponential functions appear, in the matrix exponential.

If the poles are complex, then matters get more complicated.

EXAMPLE 8.3

Find the matrix exponential for the system shown in the block diagram below.

U → $\dfrac{10}{s^2 + s + 10}$ → Y

$$\text{or} \qquad \frac{d^2Y}{dt^2} + \frac{dY}{dt} + 10Y = 10U$$

Choose $X_1 = Y$ and $X_2 = dY/dt$. Then

$$\dot{X}_1 = X_2$$
$$\dot{X}_2 = -X_2 - 10X_1 + 10U$$

$$\text{or} \qquad \dot{X} = \begin{bmatrix} 0 & 1 \\ -10 & -1 \end{bmatrix} X + \begin{bmatrix} 0 \\ 1 \end{bmatrix} U$$

$$(sI - A) = \begin{bmatrix} S & -1 \\ 10 & s+1 \end{bmatrix}$$

$$(sI - A)^{-1} = \begin{bmatrix} \dfrac{s+1}{s^2+s+10} & \dfrac{1}{s^2+s+10} \\ \dfrac{-10}{s^2+s+10} & \dfrac{s}{s^2+s+10} \end{bmatrix}$$

$$e^{At} = \mathcal{L}^{-1}[(sI - A)^{-1}] \qquad [\text{Terms in } e^{-\zeta\omega_n t} \sin(\omega t + \phi)]$$

The solution is left as an exercise for the interested student.

8.2 CALCULATING THE MATRIX EXPONENTIAL USING TRUNCATED POWER SERIES

Once we have the concept of the matrix exponential as a matrix power series, we can use that concept fruitfully in calculating the response of a linear circuit to some set of initial conditions. The most obvious way to go about this is to truncate the series at some point, and use a finite series approximation for the infinite series. Let's take a very simple case like that and truncate after just two terms in the series. Will that give anything useful? Let's try it and see.

If we just use two terms, then we are claiming

$$e^{At} \approx I + At \tag{8.2.1}$$

Do this for some specific time T. Under the right conditions this can be a useful approximation to the matrix exponential. Go back to a solution method we used earlier, the Euler algorithm. We could solve the vector matrix state equations using Euler's method, and if we did, we would find

$$\begin{aligned}
\mathbf{X}(t+T) &= \mathbf{X}(t) + \dot{\mathbf{X}}(t) * T \\
&= I * \mathbf{X}(t) + A * \mathbf{X}(t) * T \\
&= (I + (A * T)) * \mathbf{X}(t)
\end{aligned} \tag{8.2.2}$$

We can see from this that Euler's method computes $X(t + T)$ by premultiplying $X(t)$ by a two-term approximation to the matrix exponential.

We have encountered this expression and this type of analysis earlier in Section 3.3. There we found that using the Euler method forced us to choose T, an order of magnitude smaller than the smallest system time constant if we wanted our computations to be reasonably accurate. We also found that there was the possibility of numerical instability when T was too large. We are in exactly the same situation here since using the truncated matrix exponential (truncated at the AT term) is equivalent to using the Euler integration algorithm.

It seems clear that we can do more accurate computations if we are willing to carry the matrix exponential series out further. What is not obvious is that this can be equivalent to using some of the other common integration algorithms. For a fuller discussion see References [4] and [5].

EXAMPLE 8.4

In Section 8.1 we have an example system with an A matrix of

$$A = \begin{bmatrix} -1 & 0 \\ 1 & -2 \end{bmatrix}$$

This system has time constants of 1 s and 0.5 s. Choose a T of 0.1 s (one-fifth of the smaller time constant). Then, approximating e^{AT} with a two-term series, $I + AT$, we have

$$I + AT = \begin{bmatrix} 1 & 0 \\ 0 & 1 \end{bmatrix} + \begin{bmatrix} -0.1 & 0 \\ 0.1 & -0.2 \end{bmatrix} = \begin{bmatrix} 0.9 & 0 \\ 0.1 & 0.8 \end{bmatrix}$$

Compare these values with the correct values computed from the theoretical expression

$$\begin{bmatrix} 0.9049 & 0 \\ 0.0856 & 0.8193 \end{bmatrix}$$

If we do more terms we find that the series converges rapidly.

For the series with terms to T^2, we have the approximate value of the matrix exponential

$$\begin{bmatrix} 0.905 & 0 \\ 0.085 & 0.82 \end{bmatrix}$$

Taking more terms, we have:

$$\begin{bmatrix} 0.9049170 & 0.000000 \\ 0.8557811 & 0.8193389 \end{bmatrix} \quad \text{for terms up to } T^4$$

and

$$\begin{bmatrix} 0.9049170 & 0.000000 \\ 0.8557811 & 0.8193389 \end{bmatrix} \quad \text{for terms up to } T^{10}$$

We can see that nothing of significance happens by adding terms beyond T^4.

There is a point at which adding further terms becomes an exercise in futility. Sooner or later, the $N!$ in the denominator of the exponential coefficient becomes large enough that the term becomes so small that it gets rounded off in any additions that occur. When that happens, there is no sense adding many more terms. Exactly when that happens depends upon the size of the elements of the A matrix. More on this later.

EXAMPLE 8.5

Consider the circuit in the figure below. This same circuit was used in Problem 3.7. The state equations are

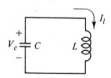

$$X(t) = \begin{bmatrix} V_c(t) \\ I_l(t) \end{bmatrix}$$

$$\dot{X}(t) = \begin{bmatrix} 0 & -1/C \\ 1/L & 0 \end{bmatrix} * X(t)$$

Pick $C = 1000 \ \mu\text{F}$ and $L = 1$ H. The natural frequency of the network is 31.6 rad/s or about 5 Hz. That gives a period of about 0.2 s. If we evaluate the matrix exponential with $DT = 0.01$ s (about 20 computations per period), we find the results given in Table 8.1.

Since the truncated series for the matrix exponential shares the numerical properties of algorithms like the Euler and Runge-Kutta, it will also share their stability properties. That implies constraints on the value of DT chosen in e^{ADT}. If the system has real poles (i.e., time-constant behavior) then we will need to

TABLE 8.1

The A matrix $=$	$\begin{bmatrix} 0.000000 & -1000.000 \\ 1.000000 & 0.000000 \end{bmatrix}$
e^{ADT} (Terms to T^2)	$\begin{bmatrix} 1.000000 & -10.00000 \\ 0.01000000 & 1.000000 \end{bmatrix}$
e^{ADT} (Terms to T^5)	$\begin{bmatrix} 0.9500347 & -9.916667 \\ 0.00991666 & 0.9500347 \end{bmatrix}$
e^{ADT} (Terms to T^{10})	$\begin{bmatrix} 0.9500347 & -9.916670 \\ 0.009916669 & 0.9500347 \end{bmatrix}$
e^{ADT} (Terms to T^{20})	$\begin{bmatrix} 0.9500347 & -9.916670 \\ 0.009916669 & 0.9500347 \end{bmatrix}$

choose DT to be at least an order of magnitude less than the smallest time constant. If the system has complex poles, then DT must be chosen to give a large number of data points in each period of the sinusoidal response. In the first example just above, we used one-fifth of the smallest time constant. In the second example, we used a DT that gave 20 computations per period. In both those cases we found that more than just two or three terms were needed in the series, but that computations with seven significant figures got into roundoff after about ten terms or so. There is some trade-off here. Using a smaller DT permits using fewer terms in the series. The two examples here show the kind of accuracy found when those conditions are not satisfied. What is not shown is the loss of stability when DT is chosen to be too small.

8.3 COMPUTING INITIAL CONDITION RESPONSE USING THE MATRIX EXPONENTIAL

If we can compute the matrix exponential for a given matrix A and time interval T, then we can compute the response of a circuit or system whenever it starts from some set of initial conditions. If the time interval is T, then it is easy to compute the response at intervals of time T, by repetitive multiplication by the matrix exponential (thus calculating the response at 0, T, $2T$, $3T$, ...).

Assume that we have some initial state vector, $X(0)$. Then, we know that $X(T)$ is given by

$$X(T) = e^{AT} * X(0) \tag{8.3.1}$$

However, we also know that $x(2T)$ is given by

$$\begin{aligned} X(2T) &= e^{2AT} * X(0) \\ &= e^{AT} * e^{AT} * X(0) \\ &= [e^{AT}]^2 * X(T) \end{aligned} \tag{8.3.2}$$

In general, $\mathbf{X}((k + 1)T)$ is given by

$$\mathbf{X}((k+1)T) = [e^{AT}] * \mathbf{X}(kT) \tag{8.3.3}$$

In this way we can get solutions for $\mathbf{x}(kT)$ using this iterative technique.

ALGORITHM

MATRIX EXPONENTIAL CALCULATION OF IC LINEAR RESPONSE

Matrix exponential techniques can be used to calculate the response of a linear system described by standard linear state equations:

$$\dot{\mathbf{X}} = A * \mathbf{X}$$

Steps in the algorithm follow.

1. Choose an initial state vector, $\mathbf{X}_0$, and a computation (integration) interval, T.
2. Compute the matrix exponential, e^{AT}, using a truncated series:

$$e_N^{AT} = \sum_{n=0}^{N} \frac{(AT)^n}{n!}$$

The analytical expression for the matrix exponential may be evaluated instead, if it is available, or can be computed.

3. Compute the system response at intervals of T seconds using the recurrence relation:

$$\mathbf{X}((K+1)T) = (e_N^{AT})\mathbf{X}(KT)$$

PROBLEM 8.2

Compute the responses of the systems below, using the indicated initial state vectors and computation intervals.

$$\dot{\mathbf{X}} = \begin{bmatrix} -1 & 0 \\ 1 & -2 \end{bmatrix} \mathbf{X} \quad \mathbf{X}(0) = \begin{bmatrix} 1 \\ 0 \end{bmatrix} \quad \mathbf{X}(0) = \begin{bmatrix} 0 \\ 1 \end{bmatrix}$$

and

$$\dot{\mathbf{X}} = \begin{bmatrix} -1 & 0 & 1 \\ 1 & -2 & 0 \\ 0 & 0 & 0 \end{bmatrix} \mathbf{X} \quad \mathbf{X}(0) = \begin{bmatrix} 0 \\ 0 \\ 1 \end{bmatrix}$$

8.4 CONVERSION TO AUTONOMOUS SYSTEMS AS A MEANS OF CALCULATING THE RESPONSE OF SYSTEMS WITH CERTAIN LAPLACE TRANSFORMABLE INPUTS

We have discussed how to calculate the response of undriven systems that start from some initial conditions. In this section we will discuss a technique that allows us to compute the response of linear time-invariant systems to steps, ramps, sinusoids and any other input that has a rational Laplace transform. We will concentrate on step, ramp, and sinusoidal input signals in this section, but indicate how to proceed for other appropriate signals.

The basic technique we want to implement involves conversion of a system with inputs into a system without inputs (an autonomous system) by adding additional states in a way that the input is generated by the new states [6]. The best way to see this is to look at an example.

EXAMPLE 8.6

Sinusoidal signals satisfy some particularly simple differential equations. Say we have a signal that is $V_{max} \sin \omega t$. That signal can be generated with the following set of differential equations:

$$\frac{dS(t)}{dt} = -\omega C(t) \qquad S(0) = 0.0$$

$$\frac{dC(t)}{dt} = \omega S(t) \qquad C(0) = 1.0$$

Now, the astute reader will recognize that $S(t)$ is the sine function, and that $C(t)$ is the cosine function. We need to take one further step and recognize that $S(t)$ and $C(t)$ can be thought of as states in a system that generates sine and cosine signals.

The general situation we want to examine will be the one shown in Figure 8.1. The input generating systems that we want to consider are all linear. (Obviously, we could use nonlinear input generating systems, but they wouldn't fit into our matrix exponential formulation.)

In the input generating system we will have a dynamic system that generates the input to our original system. Those states will be in addition to the states already in the system. We will have to append extra states to those already in the original system to account for the dynamics of generating the input. If we assume that the original system has n states, then the extra states will be X_{n+1}, $X_{n+2}, \ldots$ (as many as needed).

We will consider three specific cases that are normally important.

1. *Step inputs*

$$\text{One added state:} \qquad \frac{dX_a}{dt} = 0 \qquad\qquad (8.4.1)$$

$$X_a(0) = \text{Amplitude of step} \qquad\qquad (8.4.2)$$

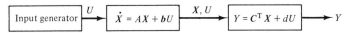

Figure 8.1 An input generating system driving a linear system.

2. *Ramp inputs*

Two added states: $\dfrac{dX_a}{dt} = 0$ (8.4.3a)

$\dfrac{dX_b}{dt} = X_a$ (8.4.3b)

$X_a(0) =$ Amplitude of ramp   (8.4.4)

3. *Sinusoidal inputs*

Two added states: $\dfrac{dX_a}{dt} = -\omega X_b$ (8.4.5a)

$\dfrac{dX_b}{dt} = +\omega X_a$ (8.4.5b)

$X_a(0) = A \sin \phi$ (8.4.6a)

$X_b(0) = A \cos \phi$ (8.4.6b)

where $A =$ Amplitude.

In all three cases, the state equation for the added state is of the form $dX/dt = A * X$, and $X(0)$ is predetermined. Also, in all three cases, the input of the original system, $u(t)$, can be taken as the first added state, X_a.

Let us consider a few examples and problems to illustrate this technique.

EXAMPLE 8.7

A second-order system with the transfer function

$$G(s) = \frac{PR}{s^2 + Qs + R}$$

has the state representation:

$$\frac{d\mathbf{X}}{dt} = \begin{bmatrix} 0 & 1 \\ -R & -Q \end{bmatrix} * \mathbf{X} + \begin{bmatrix} 0 \\ 1 \end{bmatrix} * \mathbf{U}$$
$$Y = [P \quad 0] * \mathbf{X} + 0 * \mathbf{U}$$

If we want to compute the unit step response of this system, we need to append one state to generate U. The expanded state equations are

$$\frac{d\mathbf{X}_a}{dt} = \begin{bmatrix} 0 & 1 & 0 \\ -R & -Q & 1 \\ 0 & 0 & 0 \end{bmatrix} * \mathbf{X}_a$$
$$Y = [P \quad 0 \quad 0] * \mathbf{X}_a$$
$$\mathbf{X}_a(0) = \begin{bmatrix} 0 \\ 0 \\ 1 \end{bmatrix}$$

Now we can use this representation to compute the response. We need to choose a DT for e^{ADT}. Let us assume some values for the parameters. Let $R = 1$, $Q = 0.25$, and $P = 1$. That will give a system with a natural frequency of 1 rad/s and a 0.25 damping ratio (so we'll get a little overshoot in the step response). With a natural frequency of 1, the cyclical frequency is 0.159 hz, so a period of the natural frequency is more than 6 s. Choosing a DT of 0.1 s should give halfway reasonable results, as shown in Figure 8.2.

PROBLEM 8.3

The system described in the previous example is going to be subject to inaccuracy as DT becomes larger. Experiment numerically to determine the DT at which the calculation becomes unstable when terms up to T^{10} are included in the series.

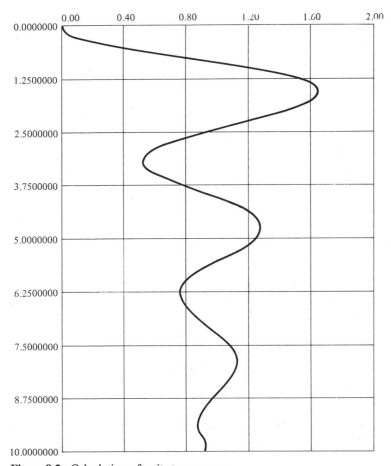

Figure 8.2 Calculation of unit step response.

PROBLEM 8.4

The DC motor is an important electrical component. The simplest model for a field-controlled motor is linear. Although that model ignores many nonlinear effects, it can be very useful.

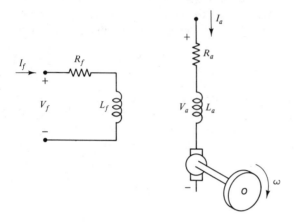

In the linear model shown, we can write a differential equation for the field current, as follows:

$$V_f = R_f I_f + L_f \frac{dI_f}{dt}$$

Then, the torque produced by the motor is proportional to the product of field current and armature current. (Two assumptions are made here. We assume that flux in the air gap is proportional to field current; that is, the relation is linear. We also assume that the armature current is constant.) The torque is applied to the mechanical load. If that load consists of an inertia, J, and a frictional force proportional to rotational velocity, we have

$$T = KI_a I_f = J \frac{d\omega}{dt} + F\omega$$

We can choose the first state to be the field current, and the second state to be rotational velocity. Then we have the state equation

$$\frac{dx_1}{dt} = \frac{dI_f}{dt} = -\frac{R_f}{L_f} I_f + \frac{V_f}{L_f}$$

$$\frac{dx_2}{dt} = \frac{d\omega}{dt} = -\frac{F}{J}\omega + \left(\frac{KI_a}{J}\right)I_f$$

(a) Using these state equations, generate the state equations for a system with a step input. Assume that a 100 volt step will produce 1000 rpm steady state.

(b) If the electrical time constant (L/R) is 0.1 s and the mechanical time constant (J/F) is 3.0 s investigate the behavior of the matrix exponential technique using *DT*s of 0.01, 0.05 and 0.1 s. Do two cases, one with 2 terms in the series, and one with 11 terms—up to $(DT)^{10}$.

8.5 ESTIMATING THE VALUE OF *DT* TO USE IN ITERATIVE COMPUTATIONS

In the last section our computations of response all required an arbitrary choice for *DT*. A better approach would be to make that choice automatically. To choose a *DT* we must first have some estimate or bound on the pole locations for the system. In this section we will examine one method for bounding the poles in the system so that we can get an upper limit for *DT*.

Factors in the characteristic polynomial of any real matrix A have one of two forms if the polynomial coefficients are real. Either we find a real pole factor of the form $(s + a)$ or we find quadratic factors (with complex poles) of the form $(s^2 + 2\zeta\omega_n s + \omega_n^2)$. If we have a real pole at $-a$, then the time constant is $1/a$, and we would probably choose *DT* to be some predetermined fraction of the time constant $1/a$—say, $1/(10a)$ or $1/(50a)$. Similarly, if we have a complex root, with a natural frequency ω_n, then we might choose *DT* to be one-twentieth of a period. That would be $(0.05)2\pi/\omega_n$ (approximately $0.3/\omega_n$).

Now, the natural frequency, ω_n and the pole "size" a have something in common. Both are the magnitude of the distance of the pole from the origin in the s plane. Figure 8.3 shows pole location in the s plane for both cases.

We can see that the value of *DT* that is used in a simulation will be determined essentially by the pole that is located farthest from the origin in the s plane. If we can get an upper limit on that maximum pole distance from the origin, then we can place a lower limit on the value of *DT* we need to use, assuming that we also know what fraction of the time constant and/or period we want to use for *DT*.

Interestingly enough, there is an easily computable upper limit on the magnitude of the largest eigenvalue of a matrix. (See problem 9.30 in Ralston and Rabinowitz [7].) The result from Ralson and Rabinowitz is that the largest eigenvalue, λ_{max} is bounded by the largest row or column sum of the absolute values of the matrix elements. More clearly,

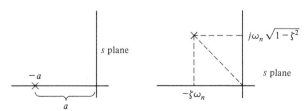

Figure 8.3 Pole locations in the s plane.

$$|\lambda_{max}| \leq MAX_i(|a_{i1}| + |a_{i2}| + \cdots + |a_{in}|) \tag{8.5.1}$$

A reasonable approach to determining DT would be

1. Bound the eigenvalues (find λ_{max}).
2. Set DT to some predetermined fraction of the reciprocal of λ_{max}.

The FORTRAN subroutine that determines DT for an arbitrary A matrix is given in Table 8.2.

In practice, MATRES (MATrix, Root EStimator) gives a conservative lower bound on time intervals, so that accuracy of simulations using this bound can be considerably better than would be expected if we took one-tenth of the fastest time constant.

EXAMPLE 8.8

Assume that we have a system with the A matrix

$$A = \begin{bmatrix} -2 & 2 \\ 0 & -1 \end{bmatrix}$$

TABLE 8.2

```
      SUBROUTINE MATRES (A,N,DT)
      DIMENSION A(20,20), R(20)
C
C MATRES PROVIDES AN UPPER BOUND ON THE RECIPROCAL
C OF THE MAXIMUM EIGENVALUE AND RETURNS THAT VALUE
C IN DT.
C
C LAST MODIFIED 11/30/84
C
         DO 10 I = 1,N
            R(I) = 0.
  10     CONTINUE
C
         DO 20 I = 1,N
            DO 19 J = 1,N
               R(I) = R(I) + ABS(A(I,J))
  19        CONTINUE
  20     CONTINUE
C
      RMAX = 0.
      DO 100 I = 1,N
         IF (R(I) .GT. RMAX) RMAX = R(I)
 100  CONTINUE
      DT = 1./RMAX
      RETURN
      END
```

Then the largest row sum is 4 ($=|-2| + |2|$). That gives a bound of 0.25 s for the lowest time constant. Actually, the lowest time constant can be found by calculating the characteristic equation of A.

$$|sI - A| = \begin{vmatrix} (s+2) & -2 \\ 0 & (s+1) \end{vmatrix} = (s+1)*(s+2)$$

Therefore the time constants associated with this A matrix are 1.0 and 0.5 s. The smallest time constant is 0.5 s, and our lower bound was computed as 0.25 s, so we are conservative by a factor of 2 in this case.

PROBLEM 8.5

A second-order system is described by the set of differential equations

$$\frac{dx_1}{dt} = Kx_2$$

$$\frac{dx_2}{dt} = -x_1 - x_2 + u$$

Determine the maximum row sum in A for a general K, and for a K of 10.0. Compare the upper bound on root size to the natural frequency for $K = 10.0$.

PROBLEM 8.6

A simple position control system is shown in the block diagram below. The gain K is adjustable, and influences the location of the system's poles. Generate a set of state equations for the system, and apply the max row sum algorithm to calculate a gain-dependent bound on the time constant. As K increases, does the time constant tend to increase or decrease? (If you have taken a course in control systems, you may have other methods for calculating the pole location and time constant. If so, compare results.)

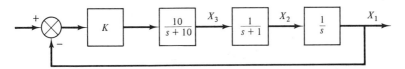

8.6 A FAST TECHNIQUE FOR COMPUTING THE RESPONSE OF SINGLE-OUTPUT SYSTEMS

There are frequently situations where we need to calculate the single output of a linear system when the input is Laplace transformable. There is a particularly fast technique that can be used in that case.

Since we have a Laplace transformable input, we will assume that the input has been subsumed into the state vector as discussed earlier. By converting to an autonomous system, we can derive a very simple difference equation for the output, $y(nT)$, of the system.

Let us assume that we have the state difference equation

$$\mathbf{X}((n+1)T) = E\mathbf{X}(nT) \qquad \text{where } E = e^{AT} \qquad (8.6.1)$$

The Cayley-Hamilton theorem [8] gives the following result.

$$P_N E^N + P_{N-1}E^{N-1} + \cdots + P_1 E + I = 0 \qquad (8.6.2)$$

In this expression, the general term is $P_j E^j$. We can premultiply the general term by C (from $y = C\mathbf{X} + D\mathbf{u}$) and postmultiply the general term by $\mathbf{X}((k-N)T)$. Then the general term becomes

$$CP_j E^j \mathbf{X}((k-N)T) = P_j y((k-N+j)T) \qquad (8.6.3)$$

Now if we do this pre- and postmultiplication to all of the terms in Eq. (8.6.1), we have

$$P_N y(kT) + P_{N-1} y((k-1)T) + \cdots + P_0 y((k-N)T) = 0 \qquad (8.6.4)$$

We can solve for $y(kT)$ in this expression, and if we can compute start-up values for y, we can use this expression to solve for $y(kT)$ repetitively.

EXAMPLE 8.9

Consider the system described by

$$\frac{d\mathbf{X}}{dt} = \begin{bmatrix} -1 & 0 \\ 1 & -2 \end{bmatrix} * \mathbf{X} \qquad \mathbf{X}(0) = \begin{bmatrix} 1 \\ 0 \end{bmatrix}$$

We have looked at this same system in examples in Sections 8.1 and 8.2. We have a theoretical expression for the matrix exponential:

$$E = e^{AT} = \begin{bmatrix} e^{-t} & 0 \\ e^{-t} - e^{-2t} & e^{-2t} \end{bmatrix}$$

If we apply the Cayley-Hamilton theorem to E, we have

$$|\lambda I - E| = (\lambda - e^{-t})(\lambda - e^{-2t})$$

From this we can conclude that the output of the system will satisfy the difference equation

$$y_k = (e^{-T} + e^{-2T})y_{k-1} + e^{-3T}y_{k-2}$$

Note that we do not need to know the matrix $\mathbf{C}$ (in $y = \mathbf{C}^T * \mathbf{X}$). (Remember that this system is autonomous—no input!)

If we want to use this algorithm in a program, we need a method of calculating the coefficients in the characteristic polynomial in an arbitrary square matrix. One method is Bocher's method [9, p. 432]:

$$P_N E^N + P_{N-1} E^{N-1} + \cdots + P_1 E + P_0 I = 0 \qquad (8.6.5)$$

Here, we will assume that $P(N)$ is 1. This causes no loss of generality. The rest of the coefficients can then be computed using

$$P_{N-1} = -T_1 \tag{8.6.6a}$$

$$P_{N-2} = -\frac{1}{2}[P_{N-1}T_1 + T_2] \tag{8.6.6b}$$

$$P_{N-3} = -\frac{1}{3}[P_{N-2}T_1 + P_{N-1}T_2 + T_3] \tag{8.6.6c}$$

$$\vdots$$

$$P_0 = -\frac{1}{N}[P_1T_1 + P_2T_2 + \cdots + P_{N-1}T_{N-1} + T_N] \tag{8.6.6d}$$

Here, T_k is the trace of the kth power of E (that is, E^k). This is not a particularly difficult routine to program, and programs are given in Appendix 2 (MATPOL and MATTRC).

Finding the characteristic polynomial coefficients is only part of the problem. The other part is to calculate start-up values of $y(t)$. However, we can calculate as many samples of $v(t)$ as we need by using the "standard" matrix exponential methods. If we have the initial state, $X(0)$, then

$$Y(0) = C^TX(0) \tag{8.6.7a}$$
$$Y(DT) = C^TX(DT) = C^TEX(0) \tag{8.6.7b}$$

$$\vdots$$

$$Y(NDT) = C^TE^NX(0) \tag{8.6.7c}$$

Once we have sufficient start-up values, we can then begin using the difference equation to compute further values for $Y(k\,DT)$.

ALGORITHM

STEP, RAMP, AND SINUSOIDAL DIFFERENCE EQUATION

To compute the step, ramp or sinusoidal response of a linear, time-invariant system develop and solve a difference equation for the system as in the steps below. Note that a different difference equation is developed for each input.

1. Assume a system description of the form

$$\dot{X} = AX + bU \qquad X(0) = 0$$
$$Y = C^TX \qquad\qquad T = \text{Calculation interval}$$

2. Assume that the input is either a step, ramp or sinusoid, and form an augmented system as described in Section 8.4. Add states

$$\dot{X} = 0 \qquad X(0) = 0 \quad \text{(Step)} \qquad \dot{X}_{in} = \begin{bmatrix} 0 & 1 \\ 0 & 0 \end{bmatrix} X_{in}$$

$$X(0) = \begin{bmatrix} 0 \\ 1 \end{bmatrix} \quad \text{(Ramp)}$$

$$\dot{X}_{in} = \begin{bmatrix} 0 & -\omega \\ \omega & 0 \end{bmatrix} X_{in} \qquad X_{in}(0) = \begin{bmatrix} 0 \\ 1 \end{bmatrix} \qquad X_1(t) = \sin \omega t \quad X_2(t) = \cos \omega t$$

In general,

$$\dot{X}_{in} = A_{in}X_{in} \qquad U = C_{in}^T X_{in}$$

$$\dot{X}_a = \begin{bmatrix} \dot{X} \\ ---- \\ \dot{X} \end{bmatrix} = \begin{bmatrix} A & | & BC_{in}^T \\ ---|--- \\ 0 & | & A_{in} \end{bmatrix} \begin{bmatrix} X \\ ---- \\ X_{in} \end{bmatrix} = A_a X_a \qquad Y = C_a^T X_a$$

3. Estimate a time interval T_{max} for computation of the matrix exponential, using the algorithm outlined in Section 8.5:

$$T_{max} = \frac{1}{(\text{MAX}_i(\Sigma_j|A_{aij}|))}$$

4. Compute the matrix exponential for the augmented system for a time interval that is (a) less than or equal to T_{max}, and (b) T divided by an integer. (Select the smallest integer that, when divided into T, produces a result less than or equal to T_{max}. Call that integer N_{pow}.
5. Raise the matrix exponential to the power N_{pow}:

$$E = (e^{A_a T_{max}}_{N_{terms}})^{N_{pow}} \qquad N_{pow} \geq \frac{T}{T_{max}}$$

6. Compute the characteristic polynomial of the matrix exponential:

$$|\lambda I - E| = p_n \lambda^n + p_{n-1}\lambda^{n-1} + \cdots + p_1\lambda + p_0$$

7. Compute N_a starting values of the output:

$$X_a((k+1)T) = EX_a(kT) \qquad k = 0, 1, \ldots N_a$$
$$Y(kT) = C_a^T X_a$$

8. Step the computation along using the difference equation:

$$Y(kT) = \frac{-1}{p_n}[p_{n-1}Y((k-1)T) + \cdots + p_0 Y((k-N_a)T)]$$

PROBLEM 8.7

Develop the difference equation for the step response of the first-order system shown.

PROBLEM 8.8

Develop the difference equation for the response to a sine input of unit amplitude for the first order system of Problem 8.7. Calculate the sine response at intervals of 0.1 s for a 0.5-Hz input signal.

8.7 COMPUTING THE RESPONSE OF LINEAR SYSTEMS TO ARBITRARY INPUTS

In many situations we need to calculate the response of a linear system to some arbitrary input. For example, we might store a record of typical inputs in a computer file and use that record to drive our simulation and computational model. So far in this chapter we have not considered this kind of situation, rather emphasizing computation of the response of autonomous systems. In this section we will focus more on the response of systems that have some arbitrary input.

The case of an arbitrary input can be most easily visualized by first examining a system with one state. Then we have

$$\frac{dx}{dt} = ax + bu \tag{8.7.1}$$

The response of this system is

$$x(t) = e^{at}x(0) + \int_0^t e^{a(t-\tau)}bu(\tau)\, d\tau \tag{8.7.2}$$

This expression consists of two terms. The first term is due to the presence of initial conditions, and is usually referred to as the zero-input response. It is the only response exhibited by the system when there is no input, and the system has some initial condition, $x(0)$. The second term is due solely to the input and is the forced response of the system. The forced response is in the familiar form of a convolution integral.

Now, imagine that we are doing a sequential computation. We have computed $x(kT)$ and we want to compute $x((k + 1)T)$. In that case, we can modify Eq. (8.7.1) to

$$x((k+1)T) = e^{aT}x(kT) + \int_{kT}^{(k+1)T} e^{a((k+1)T-\tau)}bu(\tau)\, d\tau \tag{8.7.3}$$

From our work in the first part of this chapter, we know how to compute the zero-input response using (matrix) exponential methods. We need to worry about calculating the convolution integral term.

The convolution integral term could be difficult to evaluate for general inputs. However, we are frequently dealing with inputs that are only described at the sample instants, kT. In that case we can make a few assumptions or suppositions about what happens to the input between sample instants. The simplest situation we could have would be if the input could be assumed to be constant during an entire sample period, as shown in Figure 8.4. With this assumption we can calculate the forced response as

$$x((k+1)T) = e^{aT}x(kT) + u(kT)\int_0^T e^{a(T-\tau)}b\, d\tau \tag{8.7.4}$$

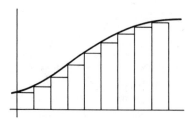

Figure 8.4 Simple calculation of integral (area).

Thus, if we can evaluate the integral of the exponential, we will have a constant multiplier for $u(kT)$ to apply in calculating $x((k + 1)T)$.

It should be obvious that this same development will carry over to the case of multiple states. In that case, the development is

$$\mathbf{X}(t) = e^{At}\mathbf{X}(0) + \int_0^t e^{A(t-\tau)}\mathbf{b}u(\tau)\,d\tau \tag{8.7.5}$$

or
$$\mathbf{X}((k+1)T) = e^{AT}\mathbf{X}(kT) + \int_{kT}^{(k+1)T} e^{A((k+1)T-\tau)}\mathbf{b}u(\tau)\,d\tau$$

$$= e^{AT}\mathbf{X}(kT) + u(kT)\int_0^T e^{A(T-\tau)}\mathbf{b}\,d\tau$$

$$= e^{AT}\mathbf{X}(kT) + \mathbf{G}u(kT) \tag{8.7.6}$$

The final expression for $\mathbf{X}((k + 1)T)$ involves the matrix exponential (evaluated for $t = T$) and a matrix $\mathbf{G}$, which is just the states produced by a unit step input. (The integral is evaluated as though $u(t)$ is 1.) If the matrix exponential and the matrix $\mathbf{G}$ can be evaluated (and we already have that problem solved), then the computation for $\mathbf{X}(kT)$ can proceed using new inputs as they occur.

To use this result requires a sequence of unknown, possibly random inputs. We might record such inputs in a file or an array, or we could be running our computer simulation and using input data as it is generated. In either case this algorithm can be used to compute response, and accuracy is limited by accuracy of computation of the matrices involved and by how well the system satisfies the assumption of constant input over the computation interval.

8.8 NUMERICAL INVERSION OF RATIONAL LAPLACE TRANSFORMS

The concepts developed in this chapter can be used for numerical inversion of rational Laplace Transform functions. If the Laplace Transform function is given by

$$F(s) = \frac{N(s)}{D(s)} \tag{8.8.1}$$

where $N(s)$ is the numerator polynomial of degree m and $D(s)$ is the denominator polynomial of degree n, then $F(s)$ can be thought of as the response of a linear

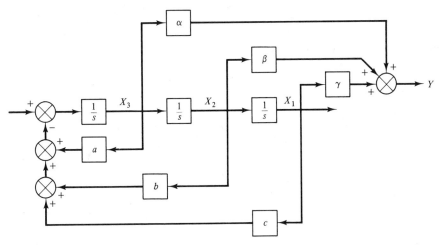

Figure 8.5 Impulse response generator.

system with appropriate initial conditions. We can also think of $F(s)$ as the impulse response of a linear, time-invariant system.

Let us pursue the approach of computing the impulse response of a system that has $F(s)$ for a transfer function. Figure 8.5 shows a system that has the impulse response for

$$F(s) = \frac{\alpha s^2 + \beta s + \gamma}{s^3 + as^2 + bs + c}$$

The example shown in Figure 8.6 is for a third-order system, but the concept can be generalized easily to systems of other orders.

For this example a set of canonical state equations can be generated. We have

$$\dot{X} = \begin{bmatrix} 0 & 1 & 0 \\ 0 & 0 & 1 \\ -c & -b & -a \end{bmatrix} X + \begin{bmatrix} 0 \\ 0 \\ 1 \end{bmatrix} U \qquad (8.8.2a)$$

$$Y = [\gamma \quad \beta \quad \alpha]X \qquad (8.8.2b)$$

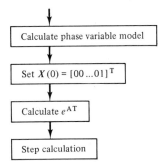

Figure 8.6 Flowchart for impulse response calculation.

If the input to this system is a unit impulse (Dirac delta function), then at $t = 0+$ only X_3 will have been changed (from zero to one). After the impulse is finished, the system will behave as though it is an autonomous system starting with the initial condition $[0 \quad 0 \quad 1]^T$.

It is easy to generalize this technique to systems of other orders, and a flowchart for the calculations is given in Figure 8.7.

If we can compute the impulse response of a linear, time-invariant system then essentially we are inverting the transfer function, converting the transfer function into its Laplace transform inverse, the impulse response. Therefore we have a technique for inverting any rational Laplace transform with denominator degree higher than the numerator degree.

REFERENCES

1. P. M. DeRusso, R. J. Roy, and C. M. Close, *State Variables for Engineers,* Wiley, New York, 1965.
2. R. C. Dorf, *Modern Control Systems,* 3rd ed., Addison-Wesley, Reading, MA, 1980.
3. B. S. Patten, (Ed), *Systems Analysis and Simulation in Ecology,* Vol. I, Academic Press, New York, 1971. (Patten's book is a compilation of articles on a number of different simulation topics. His article, "A Primer for Ecological Modeling and Simulation with Analog and Digital Computers" includes a section on matrix exponential methods, and also contains several interesting ecological systems described with linear system models.)
4. E. J. Mastascusa, A relation between Liou's method and the fourth-order Runge-Kutta method for evaluating transient response, *Proceedings of the IEEE,* vol. 57, pp. 803–804, May 1969.
5. D. E. Whitney, More about similarities between Runge-Kutta and matrix exponential methods for evaluating transient response, *Proceedings of the IEEE* (Letters), vol. 57, pp. 2053–2054, November 1969.
6. E. J. Mastascusa and J. G. Simes, Conversion to autonomous systems: A means of evaluation of transient response in linear systems, *Proceedings of the IEEE* (Letters), vol. 59, pp. 1538–1539, October, 1971.
7. A. Ralston and P. Rabinowitz, *A First Course in Numerical Analysis,* 2nd ed., McGraw-Hill, New York, 1978. (A classic text in numerical analysis.)
8. S. W. Director and R. A. Rohrer, *Introduction to Systems Theory,* McGraw-Hill, New York, 1972. (This text gives the Cayley-Hamilton theorem, as well as many other results useful in linear system analysis, all in a form that is usually readable by a junior or senior EE student.)
9. J. H. Milsum, *Biological Control Systems Analysis,* McGraw-Hill, New York, 1966. (This book should convince many readers that system theory is useful to others besides electrical engineers.)
10. P. E. Chase, Control theory and the nuclear arms race, *Bendix Technical Journal,* Autumn 1968, pp. 43–54.
11. L. F. Richardson, *Arms and Insecurity,* Boxwood Press, Pittsburgh, 1960.

MINIPROJECTS

1. Patten [3] presents an example of an ecological system that can be put into the frame-work of a linear system driven by a step input. The system consists of five states:

$$x_1 = \text{Producers}$$
$$x_2 = \text{Herbivores}$$
$$x_3 = \text{Carnivores}$$
$$x_4 = \text{Top carnivores}$$
$$x_5 = \text{Decomposers}$$

The producers are plant species and they are driven by an outside input, solar energy, and through photosynthesis they produce food used by other organisms in the food chain. The herbivores (plant eaters) feed on the producers, causing a loss of producers and a gain in the mass of the herbivores. The herbivores are also driven by an external input. People put out bread to attract them so they can watch them. Carnivores eat herbivores and top carnivores eat anything they can catch. Ultimately, all are taken care of by the decomposers.

All of the states are really energy totals in kilocalories. The differential equations for the states are given by

$$\frac{dx_1}{dt} = -6.08x_1 + 2500$$

$$\frac{dx_2}{dt} = 0.84x_1 - 15.78x_2 + 500$$

$$\frac{dx_3}{dt} = 1.79x_2 - 6.179x_3$$

$$\frac{dx_4}{dt} = 0.339x_3 - 2.142x_4$$

$$\frac{dx_5}{dt} = 1.01x_1 + 5.13x_2 + 0.74x_3 + 0.676x_4 - 188.6x_5$$

The only inputs to this system are constants (2500 and 500) in the equations for x_1 and x_2. Put this set of equations into an autonomous form. Then, using matrix exponential techniques, solve for the response of the system using the following initial conditions.

$$X(0) = (3421.26)$$
$$(\ 213.44)$$
$$(\ \ 60.06)$$
$$(\ \ \ 8.87)$$
$$(\ \ 24.38)$$

Compare your results with the results in Patten if you have access to the book.

2. There sometimes arises a situation in which you have to simulate a time delay with a finite order system. One scheme for approximating a time delay is to use the Padé approximations [9]. Two of these transfer functions are

$$P_1(s) = \frac{1 - T/2s}{1 + T/2s}$$

$$P_2(s) = \frac{1 - T/2s + T_s^2/12}{1 + T/2s + Ts^2/12}$$

(a) Form the state equations for these systems. Note that in the output equation $(Y = C^T X + DU)$ the scalar D will be nonzero.

(b) Use matrix exponential techniques to compute the step response for both systems.

3. A number of people have attempted to model the arms race between the United States and the Soviet Union. This miniproject is based on the description found in a paper by Chase [10], which appears to be based on a book by Richardson [11].

Two antagonistic countries, mutually armed, have mutually determined that neither country shall gain an arms advantage over the other. Through various nefarious intelligence activities, each country monitors the arms production and purchases of the other to estimate the total arms level of its opponent.

With an estimate of arms level available, each country then attempts to control its arms growth in such a way that it will achieve parity with its opponent in some reasonable time. To do this, each country forms an "error," that is the difference between its estimate of its opponents arms level and its own arms level. Then each country adjusts its own production rate so that if its opponent did nothing in the interim it would catch up in 6 months, assuming that the initial rate held constant. In actuality, as time goes on, the error changes, but the rate is always adjusted using that algorithm.

Simulate this system, and determine whether it is stable. Using the results of your analysis, is a bomb shelter a good investment?

Next, we have to recognize that you just can't crank up industrial production immediately. Modify the system to add a 1-month time constant in the industrial production system, so that the actual arms production rate differs from the commanded arms production rate through a system having 1-month-time constant. Answer the same questions as above for the modified system.

Chapter 9

Transfer Function and Control System Design Calculations

9.1 INTRODUCTION

Many of the techniques we have presented in this book are useful in general control system analysis. With a few additional topics we could build a useful control system analysis and design package. Working with state equation descriptions of systems we have shown how to calculate the time response of a system. However, there is a great deal of work that is done in the transfer function/frequency response domain.

In this chapter we will investigate a number of topics relating to frequency response descriptions of systems. We will look at methods for getting transfer functions from state equations, and vice versa. We will also examine methods of computing the factors (poles and zeroes) of the transfer function polynomials. Finally, we will attempt to put together many of the topics covered in earlier chapters, synthesizing methods for doing many of the calculations useful in control systems design. This area is rich in applications of the concepts and algorithms presented in this book, and examples of application of these algorithms to control system calculations will increase our appreciation of the wide usefulness of them.

9.2 OBTAINING TRANSFER FUNCTIONS FROM STATE EQUATIONS

Assume that we have a system described by a set of vector matrix state equations as follows:

$$\frac{d\mathbf{X}}{dt} = A * \mathbf{X}(t) + \mathbf{B} * U(t) \tag{9.2.1}$$

$$Y(t) = \mathbf{C}^{\mathrm{T}} * \mathbf{X}(t) + D * U(t) \tag{9.2.2}$$

Normally, when written in this form these equations assume multiple inputs and output. Now focus on the situation in which there is only a single input, $u(t)$, and a single output, $y(t)$. We are concerned with finding the transfer function that relates the transform of the output (that is, $Y(s)$) to the transform of the input (that is, $U(s)$). If Eqs. (9.2.1a) and (9.2.1b) are Laplace transformed, we may then solve for $G(s) = Y(s)/U(s)$.

Assume zero initial conditions to find the transfer function:

$$sI * \mathbf{X}(s) = A * \mathbf{X}(s) + \mathbf{B} * U(s) \tag{9.2.3}$$
$$Y(s) = \mathbf{C}^{\mathrm{T}} * \mathbf{X}(s) + D * U(s) \tag{9.2.4}$$

Solve Eq. (9.2.3) for the transform of the state vector, $X(s)$:

$$\mathbf{X}(s) = (sI - A)^{-1} * \mathbf{B} * U(s) \tag{9.2.5}$$

Now, substitute Eq. (9.2.5) in Eq. (9.2.4) to obtain

$$Y(s) = \mathbf{C}^{\mathrm{T}} * (sI - A)^{-1} * \mathbf{B} * U(s) + D * U(s) \tag{9.2.6}$$

Then the transfer function is given by

$$G(s) = \frac{Y(s)}{U(s)} = \mathbf{C}^{\mathrm{T}} * (sI - A)^{-1} * \mathbf{B} + D \tag{9.2.7}$$

Examining Eq. (9.2.7), it seems clear that the major problem in computing $G(s)$ will be in finding the inverse of $(sI - A)$. This matrix has the variable s embedded within it, and the inverse will also be a function of s.

Fortunately, there are algorithms in existence that allow a relatively straightforward computation of the transfer function, including in the process inversion of $(sI - A)$. Most of those algorithms are variations of a method found buried in a paper by Morgan [1], the Leverrier algorithm. (Morgan's paper mentions the algorithm only incidentally.) In turn, the algorithm Morgan presents is a variant of Bocher's formula, presented by DeRusso, Roy, and Close [2].

The procedure for calculating the transfer function is as follows: We start by taking the trace of the matrix A. (The trace of a square matrix is just the algebraic sum of the diagonal elements.) Denote the trace of A by $\mathrm{tr}(A)$. Then we have

$$\begin{aligned} A_1 &= A & h_1 &= \mathrm{tr}(A_1) & R_1 &= A_1 - h_1 I \tag{9.2.8}\\ A_2 &= AR_1 & h_2 &= \tfrac{1}{2}\,\mathrm{tr}(A_2) & R_2 &= A_2 - h_2 I \tag{9.2.9}\\ &\;\;\vdots & &\;\;\vdots & &\;\;\vdots \\ A_n &= AR_{n-1} & h_n &= \frac{1}{n}\mathrm{tr}(A_n) & R_n &= A_n - h_n I = 0 \tag{9.2.10} \end{aligned}$$

where

$$[sI - A]^{-1} = \frac{1}{g(s)} R(s) \tag{9.2.11}$$

and
$$g(s) = s^n - h_1 s^{n-1} - \cdots - h_n = 1sI - A1 \tag{9.2.12}$$
$$R(s) = Is^{n-1} + R_1 s^{n-2} + \cdots + R_{n-1} \tag{9.2.13}$$

This algorithm is straightforward and easily programmed. Subroutine TRASTA given in Appendix J is an implementation of this method.

Before leaving this method, it would be misleading to imply that it is without controversy. Bosley *et al.* [3] note that several of the methods for computing transfer functions have numerical problems. Some perusal of some of the references could prove educational and give a better understanding of the problems that can arise.

ALGORITHM

LEVERRIER ALGORITHM

The Leverrier algorithm forms the basis of a method to calculate the transfer function of a system described by a set of linear state equations in standard format

$$\dot{X} = AX + bU$$
$$y = C^T X + dU$$

Steps in the algorithm are

1. Assume a system described as above.
2. Form starting values:

$$A_1 = A \quad h_1 = \text{tr}(A_1) \quad R_1 = A_1 - h_1 I$$

3. Sequence through $N - 1$ more iterations:

$$A_k = AR_{k-1} \quad h_k = \left(\frac{1}{k}\right) \text{tr}(A_k)$$
$$R_k = A_k - h_k I \quad k = 2, 3, \ldots, N$$

   Note $R_N = 0$.
4. Compute $(sI - A)^{-1}$ as follows:

$$[sI - A]^{-1} = \frac{R(s)}{g(s)}$$
$$g(s) = s^n - h_1 s^{n-1} - \cdots - h_n$$
$$R(s) = Is^{n-1} + R_1 s^{n-2} + \cdots + R_{n-1} s^0.$$

5. Compute the transfer function as

$$G(s) = C^T [sI - A]^{-1} b + d$$

EXAMPLE 9.1

We use the Leverrier algorithm to find the transfer function of this system.

$$\dot{X} = 10X_2 \quad \dot{X}_2 = -X_2 - X_1 + U \quad Y = X_1$$
$$A = \begin{bmatrix} 0 & 10 \\ -1 & -1 \end{bmatrix} \quad b = \begin{bmatrix} 0 \\ 1 \end{bmatrix} \quad C = \begin{bmatrix} 1 \\ 0 \end{bmatrix}, \quad d = 0$$

Form the starting values of matrices as follows:

$$A_1 = A = \begin{bmatrix} 0 & 10 \\ -1 & -1 \end{bmatrix} \quad h_1 = -1$$

$$R_1 = \begin{bmatrix} 0 & 10 \\ -1 & -1 \end{bmatrix} + \begin{bmatrix} 1 & 0 \\ 0 & 1 \end{bmatrix} = \begin{bmatrix} 1 & 10 \\ -1 & 0 \end{bmatrix}$$

Next, sequence through all of the steps:

$$A_2 = \begin{bmatrix} 0 & 10 \\ -1 & -1 \end{bmatrix} \begin{bmatrix} 1 & 10 \\ -1 & 0 \end{bmatrix} = \begin{bmatrix} -10 & 0 \\ 0 & -10 \end{bmatrix}$$

$$h_2 = \frac{-20}{2} = -10$$

$$R_2 = \begin{bmatrix} 0 & 0 \\ 0 & 0 \end{bmatrix}$$

Finally, compute the transfer function:

$$R(s) = \begin{bmatrix} s & 0 \\ 0 & s \end{bmatrix} + \begin{bmatrix} 1 & 10 \\ -1 & 0 \end{bmatrix} = \begin{bmatrix} s+1 & 10 \\ -1 & s \end{bmatrix}$$

$$g(s) = s^2 + s + 10$$

$$G(s) = \begin{bmatrix} 1 & 0 \end{bmatrix} \begin{bmatrix} s+1 & 10 \\ -1 & s \end{bmatrix} \begin{bmatrix} 0 \\ 1 \end{bmatrix} * \frac{1}{s^2 + s + 10}$$

$$= \begin{bmatrix} 1 & 0 \end{bmatrix} \begin{bmatrix} 10 \\ s \end{bmatrix} * \frac{1}{s^2 + s + 10} = \frac{10}{s^2 + s + 10}$$

PROBLEM 9.1

Using the algorithm presented above, find the transfer functions of these two systems.

System A: $\dot{X} = \begin{bmatrix} -1 & 0 \\ 0 & -2 \end{bmatrix} X + \begin{bmatrix} 1 \\ -1 \end{bmatrix} U \quad Y = \begin{bmatrix} 1 & 1 \end{bmatrix} X$

System B: $\dot{X} = \begin{bmatrix} 1 & 0 \\ -2 & -3 \end{bmatrix} X + \begin{bmatrix} 0 \\ 1 \end{bmatrix} U \quad Y = \begin{bmatrix} 1 & 0 \end{bmatrix} X$

The Leverrier algorithm can also be used to find transfer functions of sampled systems. Consider the system described by the discrete state equations

$$\mathbf{X}(k+1) = A * \mathbf{X}(k) + \mathbf{B} * U(k) \tag{9.2.14}$$
$$Y(k) = \mathbf{C}^T * \mathbf{X}(k) + D * U(k) \tag{9.2.15}$$

Then, by taking Z transforms of both equations, with a single input and a single output, we would have

$$\frac{Y[z]}{U[z]} = \mathbf{C}^T * (zI - A)^{-1} * \mathbf{B} + D \tag{9.2.16}$$

The expression for the transfer function $Y[z]/U[z]$ is precisely the same as that for the transfer function of the continuous system given by Eq. (9.2.7) except that z appears in place of s. However, the algorithm "dosen't care" what the variable is, and we can use precisely the same algorithm to calculate the transfer function of this kind of sampled system.

PROBLEM 9.2
   Using the Leverrier algorithm, find the transfer function of these two systems:

System A: $\quad \mathbf{X}(k+1) = \begin{bmatrix} -1 & 0 \\ 0 & -2 \end{bmatrix} \mathbf{X}(k) + \begin{bmatrix} 1 \\ 1 \end{bmatrix} U(k) \quad Y = [1 \quad 1]\mathbf{X}$

System B: $\quad \mathbf{X}(k+1) = \begin{bmatrix} 1 & 0 \\ -2 & -3 \end{bmatrix} \mathbf{X}(k) + \begin{bmatrix} 0 \\ 1 \end{bmatrix} U(k) \quad Y = [1 \quad 0]\mathbf{X}$

9.3 OBTAINING STATE EQUATIONS FROM TRANSFER FUNCTIONS

In the preceding section we examined the problem of calculating numerical transfer functions from numerical state equations. It is simpler to build the bridge back from transfer functions to state equations, and we will present a short discussion of that process in order to cover all of the manipulations and changes of representations that are useful.

   For a given transfer function, there are numerous ways we can generate state equations. The state equations we eventually find will not be unique. However, there are two relatively standard approaches to generating state equations from transfer functions.

   The first way to generate state equations from transfer functions assumes that the transfer function denominator is factored. Then a partial fraction expansion of the transfer function is performed, leading to a representation that can easily be cast into a state variable representation. This method is best illustrated by an example.

EXAMPLE 9.2
   Assume that a system has the transfer function

$$G(s) = \frac{1}{(s+1)(s+2)} = \frac{1}{(s+1)} + \frac{-1}{(s+2)}$$

The partial fraction expansion of the transfer function (shown to the right) leads to the block diagram in the figure below.

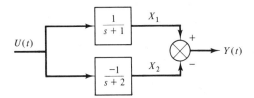

From the block diagram we could generate a set of two state equations:

$$\frac{dX_1}{dt} = -X_1 + U$$

$$\frac{dX_2}{dt} = -2X_2 - U$$

There are several problems encountered when trying to generalize this approach. First, just to be able to use partial fractions requires being able to factor the denominator polynomial. As we will see later, that involves problems of computational time, inaccuracy, and possibly unreliability. Second, even if the polynomial is correctly factored, we would have a problem with complex poles. Either we would have to do complex arithmetic or we would have to make special provision for second-order factors with less than unity damping (complex poles).

This "modal" approach may be desirable in some systems. There will be times when we want to separate out the effects of different time constants. However, in many cases the most important consideration is accuracy of computation of time response, and we may not want to get involved in the problems of factoring the denominator polynomial reliably.

Another approach to this problem is based on the following idea. Imagine that we have a differential equation. Say it is second order, as follows:

$$\frac{d^2y}{dt^2} + a\frac{dy}{dt} + by(t) = cU(t) \qquad (9.3.1)$$

Then, we can define states in terms of the output, $y(t)$, and the derivatives of the output. For the second-order system we would have

$$x_1 = y \quad \dot{x}_1 = \frac{dy}{dt} \qquad (9.3.2a)$$

$$x_2 = \frac{dy}{dt} \quad \dot{x}_2 = -ax_2 - bx_1 + cU(t) \qquad (9.3.2b)$$

This set of states is usually referred to as *phase variables*. The A matrix generated by this process is also referred to as the *companion matrix* of the differential equation [4].

Phase variables represent a particularly attractive way of converting a transfer function representation to a state variable representation. There is no requirement that the poles of the system be known. There are no obvious numerical difficulties, and the method is straightforward as long as the transfer function has a denominator of higher degree than the numerator.

EXAMPLE 9.3

A phase variable state model of the transfer function below can be generated:

$$G(s) = \frac{10}{s^2 + s + 10}$$

The phase variable model is

$$\dot{X} = \begin{bmatrix} 0 & 1 \\ -10 & -1 \end{bmatrix} X + \begin{bmatrix} 0 \\ 1 \end{bmatrix} U \quad Y = [1 \quad 0]X \quad \text{(two states)}$$

Our approach has to be modified if the system has zeros (that is, the numerator of the transfer function is not simply a constant). For example, the following third-order system with the transfer function $G(s)$ can be put into phase variable form following an approach found in Dorf [5].

$$G(s) = \frac{6s + 5}{s^3 + 2s^2 + 3s + 4} \tag{9.3.3}$$

The phase variable representation is

$$\dot{X} = \begin{bmatrix} 0 & 1 & 0 \\ 0 & 0 & 1 \\ -4 & -3 & -2 \end{bmatrix} X + \begin{bmatrix} 0 \\ 0 \\ 1 \end{bmatrix} U \tag{9.3.4a}$$

$$Y = [5 \quad 6 \quad 0]X \quad \text{(three states)} \tag{9.3.4b}$$

This approach depends upon using a signal flow graph model for the transfer function and showing the equivalence of this state model and the original transfer function. (See Dorf [5] for a readable presentation of the argument.) The net result is that the A matrix and b vector stay the same as they were in the system with a constant numerator. The c vector used to determine the output is filled in with the coefficients of the numerator polynomial in increasing order.

PROBLEM 9.3

Find the phase variable representation for the following two transfer functions:

$$G_A(s) = \frac{s^2 + 0.1s + 2}{s^3 + 7s^2 + 12s + 29}$$

$$G_B(s) = \frac{s^3 + s^2}{s^4 + 0.1s^3 + 0.01s^2 + 0.001s + 10^{-4}}$$

As long as the transfer function has a denominator of higher order than the numerator, the conversion to phase variables is straightforward. If the denominator and numerator order are equal, then we can still generate a phase variable description by taking out a constant term. The technique is best illustrated with an example.

EXAMPLE 9.4

Consider a transfer function with a cubic numerator and denominator.

$$G(s) = \frac{4s^3 + 3s^2 + 2s + 1}{s^3 + 4s^2 + s + 10}$$

This transfer function can be rewritten as the sum of a constant and a finite-bandwidth transfer function. (The constant will be nonzero only when the numerator degree equals or exceeds the denominator degree, and in that situation the system has an infinite bandwidth.) When numerator and denominator degrees are equal, we can find the constant by examining the high-frequency asymptotic behavior of the system, and we have

$$G(s) = \frac{4s^3 + 3s^2 + 2s + 1}{s^3 + 4s^2 + s + 10} = 4 - \left[\frac{13s^2 + 2s + 39}{s^3 + 4s^2 + s + 10} \right]$$

This now yields a phase variable set of state equations with a nonzero D term in the expression for the output.

$$\dot{X} = \begin{bmatrix} 0 & 1 & 0 \\ 0 & 0 & 1 \\ -10 & -1 & -4 \end{bmatrix} X + \begin{bmatrix} 0 \\ 0 \\ 1 \end{bmatrix} U$$
$$Y = [-39 \quad -2 \quad -13] X + 4U$$

PROBLEM 9.4

Generate phase variable representations of the transfer functions below. Indicate any transfer functions that cannot be put into phase variable representations.

(a) $\dfrac{3s^3 + 2s^2 + s}{s^3 + 2s^2 + 3s + 4}$

(b) $\dfrac{s^4}{s^3 + 3s^2 + 10s + 17}$

(c) $\dfrac{1}{s^3 + 3s^2 + 10s + 17}$

(d) $\dfrac{s^2}{s^2 + 4}$

9.4 FACTORING POLYNOMIALS

In circuit and system analysis the poles of a transfer function are found by finding the roots of the denominator polynomial, $P(s)$, of the transfer function. Factoring such a polynomial is a frequent task in circuit and system analysis, and there are many differing techniques available for accomplishing this task.

In Chapter 2 we considered the Newton-Raphson method as a way of solving the circuit equations in a nonlinear circuit. That method can also be used to find the roots of a polynomial. Say that we have a polynomial, $P(s)$, where

$$P(s) = P_0 + P_1 s + \cdots + P_n s^n \qquad (9.4.1)$$

Thus $P(s)$ is an nth-degree polynomial, and $P(s)$ will have n roots (which may or may not all be distinct).

If we have an approximation to a root, and that approximation is denoted s_{app}, then Newton's method can be used to improve that estimate:

$$s_{new} = s_{app} - \frac{P(s_{app})}{P'(s_{app})} \qquad (9.4.2)$$

Here, s_{new} is the new approximation to the root. Hopefully it is a better estimate. However, we can always iterate this process many times over, hopefully improving the estimate as iterations proceed.

When we reflect on this process, some flaws become apparent. First, we must possess some sort of a priori knowledge of the location of the root. That information may not be available. If it is not available, there is no initial root estimate to start the algorithm. Further, we have to have estimates of the location of all of the n roots of $P(s)$, since we will want to find all of the roots, not just one of them. Because of this, we need some sort of method that will give an estimate of root locations. Even if we are absolutely, totally enamored of the Newton-Raphson method, we need some other way to find starting values to "prime" the Newton-Raphson algorithm. In the next section, we will examine another algorithm, the Bernoulli method, that can be used to "prime" a Newton algorithm. That approach will allow us to take advantage of the good convergence properties of the Newton algorithm by switching over to it in the middle of the root-finding process.

EXAMPLE 9.5

Consider applying the Newton-Raphson algorithm to find the root of the polynomial:

$$P(s) = s^3 + s^2 + s + 1 \qquad s_0 = -1.1$$

Form the derivative polynomial:

$$P'(s) = 3s^2 + 2s + 1$$

Then iterate using the Newton-Raphson algorithm. One iteration is given next. Notice the rapid convergence toward the root at -1:

$$s_1 = s_0 - \frac{P(s_0)}{P'(s_0)} = -1.1 - \left(\frac{-0.221}{2.43}\right) = -1.0090535$$

9.5 THE BERNOULLI METHOD

The Bernoulli method is best illustrated using an example. Consider factoring the polynomial

$$P(s) = (s+1)(s+2) \qquad\qquad\qquad (9.5.1)$$
$$= s^2 + 3s + 2 \qquad\qquad\qquad (9.5.2)$$

We realize that we could get the roots of a quadratic easily, but we use this example as an illustration of this method.

Now, using the polynomial coefficients a difference equation can be constructed. Specifically, consider

$$x(k) = -3x(k-1) - 2x(k-2) \qquad\qquad (9.5.3)$$

Assume some sort of starting values for this difference equation. For purposes of argument, choose $x(0) = x(1) = 1$. Then from the difference equation we would have the results shown in Table 9.1.

If we look at this carefully, it looks as though the ratio of successive values is approaching the value of -2.

TABLE 9.1

k	x(k)	x(k)/x(k − 1)
0	1	
1	1	1.00000
2	−5	−5.00000
3	13	−2.60000
4	−29	−2.23077
5	61	−2.10345
6	−125	−2.04918
7	253	−2.02400
8	−509	−2.01186
9	1021	−2.00294
10	−2045	−2.00147

PROBLEM 9.5

Evaluate the ratios of several more successive values of $x(k)$. Determine whether those ratios are converging linearly or quadratically. (See the Chapter 2 discussion on the rate of convergence of different nonlinear-equation solution algorithms.)

PROBLEM 9.6

Form the difference equations to apply Bernoulli's method to the following polynomials.

(a) $s^2 + 5s + 6$
(b) $s^2 + 7s + 6$
(c) $s^3 + 6s^2 + 11s + 6$
(d) $s^3 + 8s^2 + 13s + 6$

PROBLEM 9.7

Carry out 20 steps of the Bernoulli algorithm for polynomials (a) and (b) in Problem 9.6. Compare the convergence rates of the two cases.

What is the explanation of this convergent behavior? Well, if we do a Z transform analysis of the difference equation, we will have

$$X(z) = \left(\frac{-3}{z}\right)X(z) + \left(\frac{-2}{z^2}\right)X(z) + \text{Initial condition terms} \qquad (9.5.4)$$

or $\qquad (z^2 + 3z + 2)X(z) = \text{Initial condition terms} \qquad (9.5.5)$

A general form for the solution of this difference equation consists of the roots of the polynomial taken to a power depending upon the iteration number (plus a few constants that have to be evaluated). The general form is

$$x(k) = C_1(-2)^k + C_2(-1)^k \qquad (9.5.6)$$

Not only does the solution depend on the roots of the original polynomial, we will also find one term in the solution dominating the solution—that is, the term for the largest root.

From this example the general technique for higher-degree polynomials should be fairly obvious. In general, if we have a polynomial $P(s)$, we isolate the highest power so that it will have a coefficient of unity, as we did above. Then, we solve a difference equation using the polynomial coefficients to weight past terms in the difference equation. What we would have is as follows:

If the polynomial is

$$P(s) = P_0 + P_1 s + \cdots + P_n s^n \qquad (9.5.7)$$

we set up a difference equation

$$x(k) = \frac{-P_{n-1}x(k-1) + \cdots + P_1 x(k-n+1) + P_0 x(k-n)}{P_n} \qquad (9.5.8)$$

Then we assume some starting values for the $x(k)$'s and solve the difference equation. We do enough iterations that we are sure that the largest root has become dominant in the calculations. This process will usually work reasonably well if the roots are real and have some reasonable separation. However, this process will find only the largest root. If we want any of the other roots, and we surely will, we will have to divide out the factor for the root found, and form a smaller-degree polynomial. We will put these concerns aside for now while we consider the construction of a program that implements Bernoulli's method.

Table 9.2 gives a subroutine, POLBRN, that implements Bernoulli's method. This subprogram:

Normalizes the polynomial to be factored.

Sets up and solves the difference equation from the polynomial (in the DO 100 loop).

Normalizes the solution (since difference equation solutions can grow out of bounds quickly for large-magnitude roots).

Determines whether the root is real or complex.

Computes either the single real root or the pair of complex roots.

There is a complication in all of this that is taken care of in POLBRN. There could be complex roots, leading to some sort of oscillatory behavior—either growing, steady, or decaying, but still oscillatory. We need to be able to detect complex roots when they dominate the difference equation calculations (that is, when they are the largest-magnitude roots).

Let us examine what the difference equation solution looks like when there are complex roots. Say we have

$$R_1 = a + jb \qquad \text{and} \qquad R_2 = a - jb \qquad (9.5.9)$$

Then the solution to the difference equation will have two terms, as follows:

$$x(k) = C_1(a + jb)^k + C_2(a - jb)^k \qquad (9.5.10)$$

(Note that C_1 and C_2 must be complex conjugates if the solution is to be real.)

TABLE 9.2

```
C
      SUBROUTINE POLBRN (PFAC,N,ROOT,NR,ITER)
C
C***********************************************************
C
C   POLBRN FINDS THE ROOTS OF A POLYNOMIAL USING BERNOULLI'S
C   METHOD
C   PFAC IS THE POLYNOMIAL TO BE FACTORED
C
C   A ROOT IS RETURNED IN THE COMPLEX VARIABLE, ROOTS
C   IF NR IS SET TO 1, A SINGLE , REAL ROOT WAS FOUND
C   IF NR IS SET TO 2, TWO COMPLEX ROOTS WERE FOUND
C
C   ITER = NUMBER OF ITERATIONS OF BERNOULLI'S METHOD
C
C***********************************************************
C
      DIMENSION PFAC(11)
      DIMENSION P(11),Q(11)
      COMPLEX ROOT
C
      IF (PFAC(N) .EQ. 0.) THEN
         WRITE (6,*)'ATTEMPTED TO FACTOR A POLYNOMIAL WITH ZERO'
         WRITE (6,*)'LEADING COEFFICIENT.   TRY AGAIN!'
         RETURN
      ENDIF
C
C   NORMALIZE THE POLYNOMIAL
C
      DO 2 I = 1,N
    2 P(I) = PFAC(I)/PFAC(N)
C
C   SET UP INITIAL CONDITIONS ON DIFFERENCE EQUATION
C
      DO 5 I = 1,N-1
    5 Q(I) = 1.
C
C***********************************************************
C
C   DIFFERENCE EQUATION SEGMENT
C
      DO 100 I = 1,ITER
         S = 0.
         DO 10 J = 1,N-1
         S = S - P(J)*Q(J)
   10 CONTINUE
C
            Q(N) = S
               DO 20 J = 1,N-1
               Q(J) = Q(J+1)
   20          CONTINUE
            QZERO = Q(1)
C
C   NORMALIZE TO LIMIT NUMBER SIZE
C
      IF (S .NE. 0.0) THEN
            DO 30 J = 1,N
            Q(J) = Q(J)/S
   30       CONTINUE
            QZERO = QZERO/S
      ENDIF
C
  100 CONTINUE
```

TABLE 9.2 (continued)

```
C
C***********************************************************
C
C    SEGMENT TO DETERMINE WHETHER A SINGLE REAL ROOT OR A
C    COMPLEX PAIR OF ROOTS WERE FOUND
C
     BSQUARED = (4.*(Q(1)*Q(3) - Q(2)**2)) * (QZERO*Q(2) - Q(1)**2)
     BSQUARED = BSQUARED - (QZERO*Q(3) - Q(1)*Q(2))**2
     BSQUARED = BSQUARED/((4.*(QZERO*Q(2) - Q(1)**2))**2)
     IF (BSQUARED .LT. 0.) THEN
C
C    REAL ROOT SEGMENT
C
     ROOT = Q(N-1)/Q(N-2)
     NR = 1
     ELSE
C
C    COMPLEX ROOT SEGMENT
C
     RIMAG = SQRT(BSQUARED)
     RREAL = (QZERO*Q(3)-Q(1)*Q(2))/(2.*QZERO*Q(2)-Q(1)**2)
     NR = 2
     ROOT = RREAL*(1.,0.) + RIMAG*(0.,1.)
     ENDIF
C
     RETURN
     END
C
```

With just these two roots, the generating difference equation would have been

$$x(k+2) - 2ax(k+1) + (a^2 + b^2)x(k) = 0 \qquad (9.5.11)$$

If we look at two successive sample times, we can get enough information to solve for a and b directly.

$$x(k+3) - 2ax(k+2) + (a^2 + b^2)x(k+1) = 0 \qquad (9.5.12)$$
$$x(k+2) - 2ax(k+1) + (a^2 + b^2)x(k) = 0 \qquad (9.5.13)$$

Now we can solve for a and for b^2. Doing this, we obtain

$$a = \frac{x(k+1)x(k+2) - x(k+1)x(k)}{x(k+2)x(k) - x(k+1)x(k+1)} \qquad (9.5.14)$$

and

$$b^2 = \frac{N}{D} \qquad (9.5.15)$$

where

$$N = 4[x(k+3)x(k+1) - x(k+1)x(k+1)][x(k+2)x(k) - x(k+1)x(k+1)]$$
$$- [x(k+1)x(k+2) - x(k+3)x(k)]^2 \qquad (9.5.16)$$
$$D = 4[x(k+2)x(k) - x(k+1)x(k+1)]^2 \qquad (9.5.17)$$

These expressions are implemented in POLBRN, and the expression for b^2 is evaluated to determine whether $b^2 > 0$. If so, it is assumed that a complex root exists, and a and b are calculated and returned to the calling program.

9.6 CONVERGENCE RATE OF THE BERNOULLI METHOD (REAL ROOTS)

The convergence rate of any algorithm is important. In Chapter 2 we compared the convergence rates of several different types of zero-locating algorithms. The same kinds of considerations are important in polynomial root finders, so we want to examine the rate of convergence of the Bernoulli method and determine what kind of convergence it has.

To get some estimate of rate of convergence, let us assume that we are using the Bernoulli method on a polynomial with two real roots. Then the solution of the difference equation will be

$$x(k) = C_1 R_1^k + C_2 R_2^k \tag{9.6.1}$$

Then the ratio between two successive iterates will be

$$\frac{x(k)}{x(k+1)} = \frac{C_1 R_1^{k+1} + C_2 R_2^{k+1}}{C_1 R_1^k + C_2 R_2^k} \tag{9.6.2}$$

With some algebra, we can get

$$\frac{x(k+1)}{x(k)} = \frac{R_1 + R_2(C_2/C_1)(R_2/R_1)^{k+1}}{1 + (C_2/C_1)((R_2/R_1)^k)} \tag{9.6.3}$$

Now, if we assume that $(R_2/R_1)^k \ll 1$, then

$$\frac{x(k+1)}{x(k)} \approx \left[R_1 + R_2 \left(\frac{C_2}{C_1}\right)\left(\frac{R_2}{R_1}\right)^k \right]\left[1 - \left(\frac{C_2}{C_1}\right)\left(\frac{R_2}{R_1}\right)^k \right] \tag{9.6.4}$$

Let $c = C_2/C_1$. Then we have

$$\frac{x(k+1)}{x(k)} \approx R_1 + c(R_2/R_1)^k (R_2 - R_1) + R_2 c(R_2/R_1)^{2k} \tag{9.6.5}$$

If $R_2/R_1 \ll 1$, then the last term can be neglected. That condition will be true whenever we have k large enough, assuming that R_2 is the smaller of the two roots. Then we can neglect the last term above, so that

$$\frac{x(k+1)}{x(k)} \approx R_1 + c\left(\frac{R_2}{R_1}\right)^k (R_2 - R_1) \tag{9.6.6}$$

The net result of this is that we can see that the convergence of this algorithm is such that the difference between the desired root, R_1, and the computed root, $R_1 + c(R_2/R_1)^k(R_2 - R_1)$, decays exponentially. That is linear behavior, as discussed in Chapter 2, and the Bernoulli algorithm converges linearly.

In the next section, we will reexamine the Newton algorithm, and it is of interest because of its quadratic convergence particularly.

9.7 THE NEWTON METHOD FOR FACTORING POLYNOMIALS

In many of the problems we have discussed in this text we have had to worry about accuracy and convergence. Polynomial factoring is no exception. The Newton method is one way of obtaining quadratic convergence in factoring polynomials. However, as pointed out in an earlier section, it needs some sort of starting value in order to proceed. Because of this it frequently happens that some other method is used to generate a "seed" for the Newton method.

There are a few complications that can arise when applying the Newton-Raphson algorithm to polynomial factoring. The first, minor complication is that many, if not most, polynomials of interest to electrical engineers have complex or imaginary roots. The iteration formula for the algorithm will always produce a real-valued iterate if a real seed is used. In other words, if s_{old} is the present computed value of a root, and s_{old} is real, then the next computed value for the root, s_{new}, will also be real, as long as all of the coefficients of the polynomial are real (the usual case). This can easily be seen from

$$s_{new} = s_{old} - \frac{P(s_{old})}{P'(s_{old})} \tag{9.7.1}$$

The problem of complex roots can be overcome if the seed has even a small imaginary part. That imaginary part will decay to insignificance if the root we are converging to is real. If the root is complex, the initially small imaginary part will grow and converge to the imaginary component of the root.

A problem that can be more severe is that, on occasion, we need to factor a polynomial with a multiple root. In that situation, the convergence properties of both algorithms, Bernoulli and Newton-Raphson, deteriorate significantly. (Several of the references have pertinent material; see Dennis & Schnabel [6], Burden & Faires [7], and Scraton [8].) Since true multiple roots do not occur all that often, and since the problem is one of convergence rate (and not complete lack of convergence) we will leave this topic for Miniproject 2 at the end of this chapter.

Various other schemes are used to overcome possible problems. For example, the Newton method may not converge to the root closest to the seed. When that happens, one root may be found twice or more and another root not found at all in the process. To avoid the embarrassment of finding the same root twice and missing another, one scheme is to "deflate" the polynomial by dividing out the factor due to the found root, lowering the degree of the polynomial in the process (lowering the degree by one if a real root is found, or by two if a complex root is found).

EXAMPLE 9.6

Say that we have found the root at -1 in the polynomial

$$P(s) = s^3 + 8s^2 + 13s + 6$$

Then we may divide out the root at -1 since $P(s)$ is given by

$$P(s) = (s+1)(s^2+7s+6)$$

Then we work on the new polynomial, $P_{new}(s)$, that is,

$$P_{new}(s) = s^2+7s+6$$

The process of deflation cannot take place without introducing some inaccuracy into the deflated polynomial. In turn, inaccuracy in the coefficients of the deflated polynomial will change the roots of the deflated polynomial so that they are different from the roots of the original polynomial. Further, the roots of a polynomial may be a sensitive function of the coefficients.

PROBLEM 9.8

This polynomial has a root at -10. Determine how accurately that root must be known for the Newton algorithm to converge to within 0.01 of the other root at -1. Assume a value for the root near -10, then deflate the polynomial, neglecting the remainder. Find the root of the resultant polynomial using the Newton algorithm.

$$s^4+12s^3+22s^2+21s+10$$

9.8 CALCULATING PHASE AND GAIN MARGIN

In control system design, two quantities of importance are gain and phase margin. We do not need an extensive knowledge of controls to understand the problems of computing these two quantities. In this section we will examine the problem of determining gain and phase margin numerically.

To understand the problems involved here, consider a system with the transfer function:

$$\frac{1}{s^3+0.2s^2+4s} \tag{9.8.1}$$

The Bode plot (logarithmic frequency response plot) for this transfer function is given in Figure 9.1. To a control system designer, the answers to two questions are important.

1. When the phase becomes $-180°$, what is the gain at that point?
2. When the gain becomes unity (0 dB on a Bode plot), what is the phase at that point?

We will look at the problem of computing gain margin first. Here, the gain is found for the 180° phase point. Then gain margin is defined as the negative dB gain at that point: that is

$$G_m = -20 \log_{10} |G(j\omega_{180})| \tag{9.8.2}$$

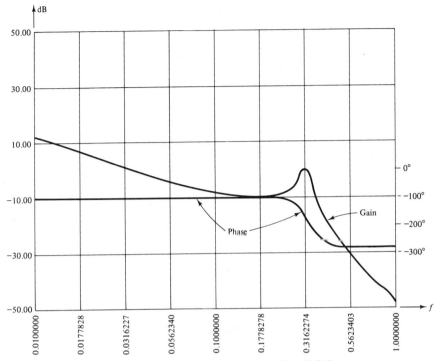

Figure 9.1 Bode plot for the transfer function given in Eq. (9.8.1).

The crucial problem is to locate the frequency for -180 degrees phase. We need to solve the equation

$$\angle G(j\omega) = -180° \qquad \text{or} \qquad f(\omega) = \angle G(j\omega) + 180° = 0 \qquad (9.8.3)$$

In Chapter 2 we discussed solving a nonlinear equation of this sort. The one-dimensional Newton algorithm could be tried, but the form of the phase function is of the sort that could produce oscillating iterations if the starting frequency guess is not close enough to the solution. For this particular problem the secant method will be more reliable.

The problem of computing phase margin presents some interesting problem variations. Essentially, we need to determine when the gain (as a function of frequency) becomes 1 (that is, 0 dB). If we plot gain against frequency, we usually find a situation like the ones shown in Figure 9.2. The general shape of these

Figure 9.2 A "typical" frequency response.

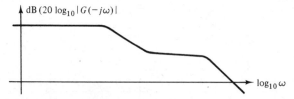

Figure 9.3 A typical Bode plot.

functions would indicate that the secant method, rather than the Newton method, should be chosen to solve this problem. However, if we think and calculate in decibels instead of pure gain, then the situation becomes different, and we have the sort of plots shown in Figure 9.3.

Conversion to decibels (Bode plots) and using logarithmic frequency produces functions that tend to straight-line segments. Those straight-line segments can produce quicker converging iterates if the Newton algorithm is used.

We leave this subject of convergence to Miniproject 4. We give a secant algorithm implementation in Appendix J, however, letting the student produce a better version if desired.

9.9 LOCATING RESONANT PEAKS IN FREQUENCY RESPONSES

Finding a resonant peak in a frequency response is essentially a maximization problem. Since the only obvious variable is frequency, finding a resonant peak is a one dimensional maximization problem. By converting to a minimization problem, we can use any of the techniques of Chapters 5 and 6 to locate the resonant peak. The Golden Section/ Fibonacci search would probably work well since frequency response plots are frequently unimodal. If multiple peaks occur in a system, unimodality can still be obtained by restricting the range of search sufficiently.

Table 9.3 shows a FORTRAN function that does the calculations to implement the resonant peak location problem as an FOFX routine suitable for use with our optimization routines.

PROBLEM 9.9

It is suspected that the following transfer function has a resonant peak at an ω somewhere betwen 3 and 4 rad/s. Do a Golden Section search to locate the resonant peak (if it is, in fact, there).

$$G(s) = \frac{s+2}{s^4 + 2s^3 + 11s^2 + 10s}$$

(a) If you are using a calculator, do ten function evaluations and determine how accurately you can locate the peak. How many function evaluations would be needed to locate the peak to within 0.01 rad/s?

TABLE 9.3

```
C
C***********************************************************
C
      SUBROUTINE F(X,FX)
      COMMON/INFO/CIRCTYPE,OUTTYPE,SIMPROB,OPTPROB,OPTALG
      CHARACTER*4 CIRCTYPE,OUTTYPE,SIMPROB,OPTPROB,OPTALG
      IF (OPTPROB .EQ. ''RESO'') CALL RESONANCE(X,FX)
      RETURN
      END
C
C***********************************************************
C
      SUBROUTINE RESONANCE (FREQ, RESP)
      DIMENSION PD(21), PN(21)
      COMPLEX POLES(20),ZEROES(20)
      COMMON/TRAN/PD,PN,ND,NN,POLES,ZEROES
      CALL TRANEVAL (FREQ,RESP,PHI,GR,GI)
      RESP = -RESP
      RETURN
      END
C
C***********************************************************
C
```

(b) If you have access to a FORTRAN system, use the function above, and determine how many function evaluations are needed to locate the peak to within 0.001 rad/s.

9.10 SUMMARY AND CONCLUSIONS

In this chapter we have shown a number of applications of techniques presented earlier applied to control system analysis. The area of control systems is not the only one rich in application of the numerical techniques we have discussed in the text, but control systems represent one of the areas richest in possible examples. (Control systems have a history of interest in numerical techniques, however. See Melsa [9] for an example of a text that incorporates a variety of numerical techniques for control system analysis.)

In general, in this text we have endeavored to present examples of algorithms for any given task. We have not given a complete presentation of all methods for any given task, and there are some areas in which there are many algorithms available for a task. One particular example is the problem of polynomial factorization, discussed in this chapter. There are many different algorithms available for this task. The Newton algorithm is very popular, but others are also widely used, including the Bairstow method [10]. Most good numerical analysis texts devote ample space to the numerous factorization methods available, and Ralston and Rabinowitz [11] give one of the more comprehensive treatments.

It should also be pointed out that we have not come close to covering all of the many applications of the algorithms developed in this text. Most of the algorithms presented have diverse applications, often far from the applications we have used for examples. Some appear disguised, appearing under other names.

Perhaps the best example of this appears in the area of adaptive filtering and identification. The Widrow-Hoff adaptive filtering algorithm is really a real-time version of a gradient optimization routine. A good review of the technique is available in Kalman and DeClaris [11] in the chapter by Widrow.

If anything, we should be careful to realize that what has been covered in this text is a bare introduction to material that, in many instances, is still being worked on actively by researchers. However, this introduction should serve well to enable you to continue to learn in the area. Good luck to you in that process.

REFERENCES

1. B. S. Morgan, Sensitivity analysis and synthesis of multivariable systems, *IEEE Transactions on Automatic Control,* vol. AC-11, July 1966, pp. 506–512.
2. P. M. DeRusso, R. J. Roy, and C. M. Close, *State Variables for Engineers,* Wiley, New York, 1965.
3. M. J. Bosley, H. W. Kropholler, F. P. Lees, and R. M. Neale, The determination of transfer functions from state variable models, *Automatica,* vol. 8, 1972, pp. 213–218. (This item gives a good description of the problems inherent in several methods of generating state models from transfer functions, and is well worth reading. There are rebuttals and discussions by Morgan and Davison that add to the interest.)
4. W-K Chen, *Linear Networks and Systems,* PWS Publishers, Boston, 1983. (See Chapter 11, especially pp. 592–600, for a good discussion of phase variables, along with several examples.)
5. R. C. Dorf, *Modern Control Systems,* Addison-Wesley, Reading, MA, 1980. (Dorf gives a very readable presentation of methods for obtaining state variable representations from signal flow graph models of transfer functions. He presents several different state models for a given transfer function, and his first model is the one most commonly used; see, particularly pp. 317–324.)
6. J. E. Dennis, Jr. and R. B. Schnabel, *Numerical Methods for Unconstrained Optimization and Nonlinear Equations,* Prentice-Hall, Englewood Cliffs, NJ, 1983.
7. R. L. Burden and J. D. Faires, *Numerical Analysis,* 3rd ed., PWS Publishers, Boston, 1985.
8. R. E. Scraton, *Basic Numerical Methods,* Edward Arnold, Ltd., London, 1984.
   References [6]–[8] discuss convergence of polynomial root-finding methods. Dennis and Schnabel give an example of linear convergence of the Newton-Raphson method in finding a double root (p. 23). Scraton also gives a slowly converging example (p. 27). Burden and Faires discuss ways of avoiding the problem (pp. 58–60). However, the method is applicable to single-equation problems (like polynomials) but not to systems of equations (as discussed in Chapter 2).
9. J. L. Melsa, *Computer Programs for Computational Assistance in the Study of Linear Control Theory,* McGraw-Hill, New York, 1970.
10. M. L. James, G. M. Smith, and J. C. Wolford, *Applied Numerical Methods for Digital Computation with FORTRAN and CSMP,* 2nd ed., IEP/Dun-Donnelley, New York, 1977.
11. A. Ralston and P. Rabinowitz, *A First Course in Numerical Analysis,* 2nd ed., McGraw-Hill, New York, 1978. (See pp. 367–397 for a very good treatment of numerous polynomial factoring methods.)

12. R. E. Kalman and N. DeClaris (Eds.), *Aspects of Network and System Theory,* Holt, Rinehart and Winston, New York, 1971.

MINIPROJECTS

1. In control systems, there is a problem encountered frequently in which we want to adjust gain of a system to meet some prespecified performance objective. Two such performance specifications are bandwidth and overshoot (in response to a step input). Frequently, in a control system there is an adjustable gain that may be used to change the behavior of the system. In a given system the gain is to be adjusted to achieve either a predetermined bandwidth or overshoot.

   (a) Determine a numerical method of measuring bandwidth (overshoot) laying out a flowchart of the process you will use.

   (b) Determine at least two alternative methods of adjusting the gain to achieve a predetermined bandwidth or overshoot, with provision for escaping from the process when an impossible situation is encountered.

2. The two polynomial factoring routines presented first in this chapter are the Bernoulli method and the Newton-Raphson method. The Bernoulli method has linear convergence for a single real root, and the Newton-Raphson method is quadratically convergent. Examine and compare the behavior of these two algorithms, both theoretically and experimentally, when a multiple root is encountered.

   (a) Examine convergence with just two poles at -1.

   (b) Generalize the two multiple (real) poles.

   (c) Examine convergence with multiple complex poles. Start with poles at $-1 + j$, $-1 + j$, $-1 - j$, $-1 - j$.

3. The Bernoulli method depends upon one root dominating the solution of a difference equation. Consider a polynomial with two roots at -1 and $-1 + a$. Here, a is a small number.

   (a) Determine the influence of a on the convergence to the root at -1. (If $a > 0$, the root at -1 is the larger root.)

   (b) Determine the result of the inaccuracy of the found root (should be at -1) on the location of the second found root.

4. In Section 9.8 we discussed how it might be possible to get better convergence in calculating phase margin by changing representation to a Bode plot form. For this project, investigate rates of convergence in calculating phase margin using the following representations.

   (a) Gain and frequency.

   (b) Decibel gain and frequency.

   (c) Decibel gain and logarithmic frequency (use $\log f$ instead of f in the search, and convert back to frequency only when the search is finished).

5. Another control system problem is determining the point at which maximum phase lead occurs in a transfer function. Say we have a transfer function, $G(s)$, in standard form.

   (a) Select an algorithm to determine the frequency at which maximum phase lead occurs.

   (b) Prepare a flowchart of the algorithm for finding maximum phase lead.

   (c) Write and debug a FORTRAN program to implement your algorithm.

Building Your Own Interactive Circuit Analysis Program

With the numerical tools presented in Chapters 1 through 3 it is possible to build a useful circuit analysis program that can compute DC levels in circuits, including circuits with nonlinear devices, do parameter variations, and do transient analyses. We want to consider some practical aspects of the construction of such a program.

To design a circuit analysis program, we might start by listing some of the characteristics that such a program will have. For our purposes here, the list will be the one below. However, the student may not agree with all of these characteristics and may well wish to add or delete something. This list and program are offered as a first attempt at developing a personal circuit analysis program, and the entire program is available from the author for anyone teaching from this text.

The circuit analysis program we can build will have at least the following characteristics.

1. The program will be interactive, not batch-oriented.
2. The program will permit the user to define a circuit, allowing the user to specify component connections, types, values and auxiliary data (initial conditions for L and C types of components, for example).
3. The program will permit the user to modify the circuit and to store and retrieve circuit descriptions.
4. The program will allow the user to specify output, including which voltage differences are to be displayed, and the program will have capability of displaying some sort of graphical output.
5. The program will do parameter variations and set parameter values to achieve specified voltages within the circuit.
6. The program will calculate transient responses of the circuit to suddenly applied voltage and current sources.

In this appendix we will consider the construction of a circuit analysis program with the features listed above. As a first step we will examine a short exercise designed to help understand the construction of an interactive program (in this case, one that does some simple vector calculations). Then after some consideration of how to represent circuit data, including the particulars of the program we will construct, we proceed to the construction of such a program. Unfortunately, a complete listing of the program cannot be given here. However, at this time the intention is to make all versions of the program available to anyone adopting this text for classroom use. At the end of this appendix a listing of the shortest version of BU-CAP can be found. Through the text and in other appendixes we will give portions of code that can be used to extend the capability of this simple version.

EXAMPLE PROBLEM: AN INTERACTIVE PROGRAM MAINLINE

An interactive mainline program is one that accepts a list of commands, rejecting undefined or illegal commands. If a legal command is entered, then the program should take whatever action is indicated by the command. After taking action the program should cycle back, prompt for another command and wait for such a command to be entered.

On the next page we have an interactive program that does various sorts of calculations with vectors. Two vectors can be defined within the "workspace," and the program keeps track of which vector(s) have been defined. The user can input two vectors (labeled 1 and 2) and can (apparently) add two vectors, form the dot product between two vectors, calculate the length of either vector, list the vectors in the workspace, and terminate execution.

The problem for the student here is to modify this program so that the following commands can be used in this program, including correcting the defects noted after the command list.

VIN Lets the user input one of two vectors.

LIST Lists the vectors defined in the workspace.

ADD Prints the result of a vector addition of V_1 and V_2.

DOT Forms the dot product of V_1 and V_2 and prints result.

LEN Calculates the length of either V_1 or V_2.

END Terminates program execution.

The defects in the program include at least the following.

1. The response to the "VIN" command should be rewritten as a subroutine.
2. The response to the "DOT" command should be rewritten as a subroutine.
3. The "ADD" command needs to be added to the system.
4. The "DOT" (and the "ADD") command need to check for vectors of different dimensions.

```
C
C EXAMPLE PROGRAM TO ILLUSTRATE INTERACTIVE PROGRAMS
C
      CHARACTER*4 REPLY
      LOGICAL V1, V2
```

```
C
      COMMON V(10,2),NDIM
      COMMON/LOGICAL/V1,V2
C
      V1 = .FALSE.
      V2 = .FALSE.
C
      WRITE(6,*)' Welcome to the Dingbat Program for Vectors!'
    1 WRITE(6,*)' Input a Vector Command.'
      READ(5,1000) REPLY
 1000 FORMAT(A4)
C
      IF (REPLY .EQ. ''VIN'') THEN
        WRITE(6,*)' Are you going to input Vector 1 or Vector 2?'
        READ(5,*) NVEC
        WRITE(6,*)' Input vector dimension.'
        READ(5,*) NDIM
        WRITE(6,*)' Input vector components.'
        READ(5,*) (V(I,NVEC), I=1,NDIM)
        WRITE(6,*)' '
        WRITE(6,*)' Your vector is:'
        WRITE(6,*) (V(I,NVEC), I=1,NDIM)
        IF (NVEC .EQ. 1) V1 = .TRUE.
        IF (NVEC .EQ. 2) V2 = .TRUE.
      ENDIF
C
      IF (REPLY .EQ. ''DOT'') CALL DOT
C
      IF (REPLY .EQ. ''ADD'') CALL ADD
C
      IF (REPLY .EQ. ''LIST'') CALL LIST
C
      IF (REPLY .EQ. ''LEN'') CALL LEN
C
      IF (REPLY .EQ. ''END'') CALL EXIT
C
      WRITE (6,*)' You have used an undefined command.'
      WRITE (6,*)' Try again.'
      WRITE (6,*)' '
      GO TO 1
      END
```

CIRCUIT REPRESENTATIONS IN CIRCUIT ANALYSIS PROGRAMS

In the design of a circuit analysis program, one very important decision that has to be made concerns how the information describing the circuit will be coded in the program. A simple circuit description in SPICE looks like the one below.

```
SPICE EXAMPLE 1
R1 1 2 1000
R2 2 3 1000
C1 2 0 1UF
C2 3 0 1UF
VIN 1 0 PULSE(0 10)
.TRAN 100USEC 4MSEC
.PRINT TRAN V(1) V(2) V(3)
.END
```

In this listing, in the first column the element is identified by element type, and in this example we have resistors (R), capacitors (C), and a voltage source (V). In the second column we find the interconnection information—that is, the nodes to which each element is connected. The first column contains the node that is "positive" and the second node column contains the number of the "negative" node. In the last column we find the numerical value of the element. We can also see that more information could be required. For example, if a capacitor is initially charged, that information would require more data space.

It is fairly easy to visualize that each kind of information will require some sort of data array in the program workspace, and that most (if not all) subroutines will need to have access to all of that information.

From the algorithms and ideas in Chapters 1 through 3 we can construct a fairly decent circuit analysis program. In the next set of material, we will define the ways in which circuit data is stored in the program we have developed.

A SIMPLE CIRCUIT ANALYSIS PROGRAM

We can construct a simple circuit analysis program using the command input structure and the circuit data structure outlined above. In this section we will examine a prototype circuit analysis program called "BU-CAP" (for Bucknell University Circuit Analysis Program). The FORTRAN source for the simplest version of BU-CAP is given later. Examine it and you will find a mainline program that has a command input structure similar to the vector program in the exercise.

In BU-CAP, each command does not automatically lead to a call of a subroutine. The bookkeeping for data entry is handled right in the mainline, and data is entered directly into the appropriate data arrays. The commands in this version are given in Table A.1. This particular version can only handle networks with resistors and sources. Even in a situation that simple many of the problems of more sophisticated situations emerge.

Most of the problems encountered in design of this sort of program center on the decisions that must be made. In this case, some of those decisions are:

* What sort of data/array structure has to be used to carry the circuit description and how should that information be shared between various program segments?

TABLE A.1 SUMMARY OF COMMANDS IN BU-CAP/VERSION 1

R, E, I   Used to input resistors (R), voltage sources (E), and current sources (I).

LIST   Used to list the circuit data within the workspace.

MDFY, ELIM   Used to modify (MDFY) a circuit description, or to eliminate (ELIM) an element from the circuit description in the workspace.

NODES   Used to set the node voltage differences that will be computed.

VTVM   Used to compute node voltage differences.

GET, SAVE   Used to save a circuit description in a file (SAVE) or to retrieve a circuit description previously saved in a file (GET).

HELP   Will print a short message about any of the commands available.

END   Used to exit the program peacefully.

* What kind of approach should be taken for solving for the desired output?

* What kinds of numerical algorithms should be chosen to calculate the solution?

```
C
C PROGRAM TO DO CIRCUIT ANALYSIS FOR NETWORKS CONTAINING R, E, I
C ELEMENTS.
C
C***********************************************************
C
      COMMON/CIRCUIT/NODES(20,4),VALUES(20),SUPP(20)/CHAR/KINDS(20)
      DIMENSION NODOUT(20,2), V(20)
      CHARACTER*1 KINDS
      CHARACTER*4 REPLY,YESORNO,FILE,FTYP
C
C NODES CONTAINS INFORMATION ON THE NODES AN ELEMENT IS CONNECTED TO:
C
C     NODES(I,1) HAS THE NUMBER OF THE POSITIVE NODE FOR ELEMENT I,
C     NODES(I,2) HAS THE NUMBER OF THE NEGATIVE NODE FOR ELEMENT I.
C
C     VALUES(I) CONTAINS THE VALUE OF ELEMENT I
C
C     KINDS(I) DETERMINES THE KIND OF ELEMENT ACCORDING TO:
C
C         KINDS(I)=E FOR A VOLTAGE SOURCE
C         KINDS(I)=I FOR A CURRENT SOURCE
C         KINDS(I)=R FOR A RESISTOR
C
C     SUPP(I) CONTAINS SUPPLEMENTARY INFORMATION FOR AN ELEMENT.
C
C         SUPP(I) = INTERNAL RESISTANCE FOR SOURCES
C
C
C***********************************************************
      WRITE (6,*)' WELCOME TO THE BUCKNELL CIRCUIT ANALYSIS PROGRAM.'
      WRITE (6,*)' VERSION 1.0, February 5, 1987'
      DO 100 J=1,5
         NODOUT(J,1) = J
         NODOUT(J,2) = 0
  100 CONTINUE
      NOUT = 5
      DO 101 J=1,4
      DO 101 I=1,20
         NODES(I,J) = 0
  101 CONTINUE
C
C
      NELMT = 0
 1000 FORMAT (A4)
    1 CONTINUE
      WRITE (6,*) '-/\/\/-'
      WRITE (6,*) ' '
C
      READ (5,1000) REPLY
C
      IF (REPLY .EQ. 'R') THEN
         NELMT = NELMT + 1
         WRITE (6,*)' NODES          RESISTANCE VALUE (KOHMS)'
         READ (5,*) NODES(NELMT,1), NODES(NELMT,2), VALUES(NELMT)
         VALUES(NELMT) = VALUES(NELMT)*1000.
         SUPP(NELMT) = 0.
         KINDS(NELMT) = REPLY
      ENDIF
      IF (REPLY .EQ. 'R') GO TO 1
```

```
C
        IF (REPLY .EQ. ''E'') THEN
           NELMT = NELMT + 1
           WRITE (6,*)' NODES     VOLTAGE     INTERNAL RESISTANCE (KOHMS)'
        READ(5,*)NODES(NELMT,1),NODES(NELMT,2),VALUES(NELMT),SUPP(NELMT)
           SUPP(NELMT) = SUPP(NELMT)*1000.
           KINDS(NELMT) = REPLY
        ENDIF
        IF (REPLY .EQ. ''E'') GO TO 1
C
        IF (REPLY .EQ. ''I'') THEN
           NELMT = NELMT + 1
           WRITE (6,*)' NODES     CURRENT'
           READ (5,*) NODES(NELMT,1), NODES(NELMT,2), VALUES(NELMT)
           KINDS(NELMT) = REPLY
        ENDIF
        IF (REPLY .EQ. ''I'') GO TO 1
C
        IF (REPLY .EQ. ''LIST'') THEN
           WRITE (6,*)' TYPE, NODES, VALUE, IC/Rint, ELEMENT #'
           DO 300 J = 1,NELMT
           WRITE (6,*) ' '
           WRITE (6,*) KINDS(J),NODES(J,1),NODES(J,2),VALUES(J),SUPP(J),J
  300      CONTINUE
        ENDIF
        IF (REPLY .EQ. ''LIST'') GO TO 1
C
        IF (REPLY .EQ. ''VTVM'') THEN
C
           CALL LINSOLVE(NELMT,NOUT,NODOUT,V)
C
           DO 500 J = 1,NOUT
           NPLUS = NODOUT(J,1)
           NMINUS = NODOUT(J,2)
              WRITE (6,*) 'V',NPLUS,'-V',NMINUS,'=',V(J)
  500      CONTINUE
        ENDIF
           WRITE (6,*) ' '
        IF (REPLY .EQ. ''VTVM'') GO TO 1
C
        IF (REPLY .EQ. ''NODE'') THEN
           WRITE (6,*)' HOW MANY NODE PAIRS ARE TO BE OUTPUT?'
           READ (5,*) NOUT
           DO 600 J = 1,NOUT
             WRITE (6,*)' PLUSNODE    MINUSNODE'
             READ (5,*) NODOUT(J,1), NODOUT(J,2)
  600      CONTINUE
           WRITE (6,*)' THANKS!'
        ENDIF
        IF (REPLY .EQ. ''NODE'') GO TO 1
C
        IF (REPLY .EQ. ''HELP'') THEN
           CALL HELP
        ENDIF
        IF (REPLY .EQ. ''HELP'') GO TO 1
C
        IF (REPLY .EQ. ''MDFY'') THEN
           WRITE(6,*)' WHICH ELEMENT DO YOU WANT TO MODIFY?'
           WRITE(6,*)' TYPE THE ELEMENT NUMBER'
           READ (5,*) NUM
           WRITE(6,*)' ELEMENT',NUM,'IS TYPE ',KINDS(NUM)
           WRITE(6,*)' DO YOU WISH TO CHANGE TYPE? (Y/N)'
           READ (5,1000) YESORNO
           IF (YESORNO) .EQ. ''Y'') THEN
              WRITE(6,*)' INPUT NEW ELEMENT TYPE.'
```

```
                READ (5,1000) KINDS(NUM)
            ENDIF
            WRITE(6,*)' DO YOU WISH TO CHANGE NODES? (Y/N)'
            READ (5,1000) YESORNO
            IF (YESORNO .EQ. ''Y'') THEN
                WRITE(6,*)' INPUT NEW +NODE AND -NODE.'
                READ (5,*) NODES(NUM,1), NODES(NUM,2)
            ENDIF
            WRITE(6,*)' DO YOU WANT TO CHANGE ELEMENT DATA? (Y/N)'
            READ (5,1000) YESORNO
            IF (YESORNO) .EQ. ''Y'') THEN
                IF(KINDS(NUM) .EQ. ''R'') THEN
                    WRITE(6,*)' INPUT R (KOHMS).'
                    READ (5,*) VALUES(NUM)
                    VALUES(NUM) = VALUES(NUM)*1000.
                ENDIF
                IF(KINDS(NUM) .EQ. ''E'') THEN
                    WRITE(6,*)' INPUT E (VOLTS) AND Rint (KOHMS).'
                    READ (5,*) VALUES(NUM), SUPP(NUM)
                    SUPP(NUM) = SUPP(NUM)*1000.
                ENDIF
                IF(KINDS(NUM) .EQ. ''I'') THEN
                    WRITE(6,*)' INPUT I (AMPS).'
                    READ (5,*) VALUES(NUM)
                ENDIF
            ENDIF
        ENDIF
        IF (REPLY .EQ. ''MDFY'') GO TO 1
C
        IF (REPLY .EQ. ''ELIM'') THEN
            WRITE(6,*)' WHICH ELEMENT DO YOU WISH TO ELIMINATE?'
            WRITE(6,*)' TYPE 0 (ZERO) IF YOU CHANGED YOUR MIND.'
            READ (5,*) NUM
            IF (NUM .NE. 0) THEN
                DO 900 J=NUM,NELMT
                    NODES(J,1)=NODES((J+1),1)
                    NODES(J,2)=NODES((J+1),2)
                    VALUES(J) = VALUES(J+1)
                    SUPP(J)   - SUPP(J+1)
                    KINDS(J)  = KINDS(J+1)
900             CONTINUE
                NELMT = NELMT - 1
            ENDIF
        ENDIF
        IF (REPLY .EQ. ''ELIM'') GO TO 1
C
C
        IF (REPLY .EQ. ''GET'') THEN
            WRITE(6,*)' Type in the name of the file to get data from.'
            READ (5,1000) FILE
            OPEN (UNIT=9, FILE=FILE)
            READ (9,1000) FTYP
            IF (FTYP .NE. '' CKT'') THEN
                WRITE(6,*)' FILE IS NOT A CIRCUIT TYPE.'
            ELSE
                READ(9,*) NELMT
                DO 950 J = 1,NELMT
                 READ(9,959) KINDS(J),NODES(J,1),NODES(J,2),NODES(J,3),
     1           NODES(J,4),VALUES(J),SUPP(J)
950             CONTINUE
            CLOSE (UNIT=9)
            ENDIF
        ENDIF
        IF (REPLY .EQ. ''GET'') GO TO 1
C
```

```
      IF (REPLY .EQ. ''SAVE'') THEN
         WRITE (6,*)' Type in the name of the file to save data in.'
         READ (5,1000) FILE
         OPEN (UNIT=9, FILE=FILE)
         WRITE (9,1000) '' CKT''
         WRITE (9,*) NELMT
         DO 960 J = 1,NELMT
            WRITE(9,959) KINDS(J),NODES(J,1),NODES(J,2),NODES(J,3),
     1        NODES(J,4),VALUES(J),SUPP(J)
  959 FORMAT (A4,4I2,2E15.6)
  960    CONTINUE
         ENDFILE (UNIT=9)
         CLOSE (UNIT=9)
      ENDIF
      IF (REPLY .EQ. ''SAVE'') GO TO 1
C
      IF (REPLY .EQ. ''END'') CALL EXIT
      WRITE(6,*)' THAT IS AN INVALID RESPONSE.  TRY AGAIN.'
      GO TO 1
      END
C
C*************************************************************************
C
      SUBROUTINE TOPOTOCOND(NELEMENTS,G,SOURCE)
C
      COMMON/CIRCUIT/NODES(20,4),VALUES(20),SUPP(20)/CHAR/KINDS(20)
      DIMENSION G(20,20), SOURCE(20)
      CHARACTER*1 KINDS
C
C ZERO CONDUCTANCE MATRIX AND SOURCE VECTOR
C
      DO 10 I=1,10
      SOURCE(I) = 0.
      DO 5 J = 1,10
      G(I,J) = 0.
    5 CONTINUE
   10 CONTINUE
C
C THIS SUBROUTINE GENERATES A CONDUCTANCE MATRIX AND SOURCE VECTOR
C FROM TOPOLOGICAL INFORMATION CONTAINED IN NODES, VALUES AND KINDS.
C
      DO 100 I=1,NELEMENTS
       IF (KINDS(I).EQ. ''R'' ) THEN
        IF(NODES(I,1).NE.0) THEN
        G(NODES(I,1),NODES(I,1))=G(NODES(I,1),NODES(I,1))+1./VALUES(I)
        IF(NODES(I,2).NE.0) THEN
         G(NODES(I,2),NODES(I,2))=G(NODES(I,2),NODES(I,2))+1./VALUES(I)
         G(NODES(I,1),NODES(I,2))=G(NODES(I,1),NODES(I,2))-1./VALUES(I)
         G(NODES(I,2),NODES(I,1))=G(NODES(I,2),NODES(I,1))-1./VALUES(I)
        ENDIF
       ELSE
        IF(NODES(I,2).NE.0) THEN
         G(NODES(I,2),NODES(I,2))=G(NODES(I,2),NODES(I,2))+1./VALUES(I)
        ENDIF
       ENDIF
      ENDIF
  100 CONTINUE
C
C NOW GENERATE THE SOURCE VECTOR
C
C
      DO 200 I = 1,NELEMENTS
C
C CHECK TO SEE IF WE HAVE FOUND A VOLTAGE SOURCE
C
```

```
          IF(KINDS(I).EQ. ''E'' ) THEN
          IF(NODES(I,1).NE.0) THEN
           G(NODES(I,1),NODES(I,1))=G(NODES(I,1),NODES(I,1))+1./SUPP(I)
           SOURCE(NODES(I,1)) = SOURCE(NODES(I,1)) + VALUES(I)/SUPP(I)
           IF(NODES(I,2).NE.0) THEN
            G(NODES(I,2),NODES(I,2))=G(NODES(I,2),NODES(I,2))+1./SUPP(I)
            G(NODES(I,1),NODES(I,2))=G(NODES(I,1),NODES(I,2))-1./SUPP(I)
            G(NODES(I,2),NODES(I,1))=G(NODES(I,2),NODES(I,1))-1./SUPP(I)
            SOURCE(NODES(I,2)) = SOURCE(NODES(I,2)) - VALUES(I)/SUPP(I)
           ENDIF
          ELSE
           IF(NODES(I,2).NE.0) THEN
            G(NODES(I,2),NODES(I,2))=G(NODES(I,2),NODES(I,2))+1./SUPP(I)
            SOURCE(NODES(I,2)) = SOURCE(NODES(I,2)) - VALUES(I)/SUPP(I)
           ENDIF
          ENDIF
         ENDIF
  200 CONTINUE
C
C NOW SEARCH FOR ALL CURRENT SOURCES
C
      DO 300 I = 1,NELEMENTS
        IF(KINDS(I).EQ. ''I'' ) THEN
          IF(NODES(I,1).NE.0) THEN
            SOURCE(NODES(I,1))=SOURCE(NODES(I,1))+VALUES(I)
          ENDIF
          IF(NODES(I,2).NE.0) THEN
            SOURCE(NODES(I,2))=SOURCE(NODES(I,2))-VALUES(I)
          ENDIF
        ENDIF
  300 CONTINUE
C
      RETURN
      END
C
C***************************************************************
C
      SUBROUTINE LINGAU (A,RHS,N)
C
C LINGAU USES GAUSSIAN REDUCTION TO SOLVE A SET OF LINEAR
C SIMULTANEOUS EQUATIONS, A*X = RHS.
C
C Version = July 24, 1985
C
      DIMENSION A(20,20), RHS(20)
C
C CHECK TO SEE IF A IS JUST 1 BY 1.
C
            IF (N .EQ. 1) THEN
                RHS(1) = RHS(1)/A(1,1)
                RETURN
            ENDIF
      DO 100 M = 2,N
      M1 = M-1
        IF(A(M1,M1) .EQ. 0.) THEN
                CALL MATPIVOT (A,RHS,M1,M1,N,N)
        ENDIF
C
C ZERO OUT COLUMN ELEMENTS BELOW DIAGONAL
C
      AM = 1./A(M1,M1)
      DO 100 K = M,N
      AK = -A(K,M1)*AM
        DO 20 L = M1,N
   20   A(K,L) = A(K,L) + AK*A(M1,L)
C
```

```
          RHS(K)  = RHS(K)  + AK*RHS(M1)
C
  100 CONTINUE
C
      N1 = N-1
      DO 300 M = 1,N1
      J = N-M+1
      AM = 1./A(J,J)
      M1 = J-1
          DO 290 K = 1,M1
          J1 = J-K
          AK = -AM*A(J1,J)
          A(J1,J) = 0.
              RHS(J1)  = RHS(J1)  + AK*RHS(J)
  290     CONTINUE
  300 CONTINUE
C
      DO 400 M = 1,N
      AM = 1./A(M,M)
  390     RHS(M)  = RHS(M) *AM
  400 CONTINUE
      RETURN
      END
C
C***************************************************************
C
C
      SUBROUTINE MATPIVOT (A,RHS,I,J,N,M)
C
C MATPIVOT SELECTS THE MAXIMUM ELEMENT
C FROM I TO N, AND J TO M
C AND PIVOTS THAT ELEMENT TO POSITION (I,J).
C
C Version = July 24, 1985
C
      DIMENSION A(20,20),RHS(20)
C
      AMAX = ABS(A(I,J))
      IMAX = I
      JMAX = J
C
      DO 10 MI=I,N
      DO 10 MJ=J,M
C
      IF (ABS(A(MI,MJ)) .GT. AMAX) THEN
          AMAX = ABS(A(MI,MJ))
          IMAX = MI
          JMAX = MJ
      ENDIF
   10 CONTINUE
C
C EXCHANGE ROWS IF NECESSARY
C
      IF (IMAX .GT. I) THEN
          DO 20 MI=1,N
              ATEMP = A(I,MI)
              A(I,MI) = A(IMAX,MI)
              A(IMAX,MI) = ATEMP
   20     CONTINUE
          RTEMP = RHS(I)
          RHS(I) = RHS(IMAX)
          RHS(IMAX) = RTEMP
      ENDIF
C
      RETURN
      END
```

```
C
C******************************************************************
C
      SUBROUTINE LINSOLVE(NELMT,NOUT,NODOUT,VSOLVE)
      COMMON/CIRCUIT/NODES(20,4),VALUES(20),SUPP(20)/CHAR/KINDS(20)
      DIMENSION G(20,20), GI(20,20), NODOUT(20,2), V(20), SOURCE(20)
      DIMENSION VSOLVE(20)
C
         CALL TOPOTOCOND(NELMT,G,SOURCE)
         CALL NODENUMBER (G,N)
         DO 500 I = 1,N
            V(I) = SOURCE(I)
  500    CONTINUE
         CALL LINGAU (G,V,N)
         DO 550 J=1,NOUT
         NPLUS = NODOUT(J,1)
         NMINUS= NODOUT(J,2)
         IF (NPLUS .EQ. 0) VSOLVE(J)=-V(NMINUS)
         IF (NMINUS.EQ. 0) VSOLVE(J)= V(NPLUS)
         IF((NPLUS.NE.0).AND.(NMINUS.NE.0)) VSOLVE(J)=V(NPLUS)-V(NMINUS)
  550    CONTINUE
      RETURN
      END
C
C******************************************************************
C
      SUBROUTINE NODENUMBER(G,N)
C
C NODENUMBER FINDS HOW MANY DIAGONAL ELEMENTS OF
C A CONDUCTANCE MATRIX G ARE NON-ZERO TO ESTIMATE
C THE NUMBER OF UNKNOWN NODES IN A CIRCUIT
C
      DIMENSION G(20,20)
      DO 100 I = 1,10
        IF(G(I,I) .LE. 0.) THEN
           N = I-1
           RETURN
        ENDIF
  100 CONTINUE
      RETURN
      END
C
C******************************************************************
C
      SUBROUTINE HELP
      CHARACTER*4 REPLY
      WRITE(6,*)' HELP IS AVAILABLE FOR THE FOLLOWING COMMANDS:'
      WRITE(6,*)' R          E       I       LIST'
      WRITE(6,*)' NODES    VTVM    MDFY    ELIM'
      WRITE(6,*)' UNITS    SAVE    GET'
      WRITE (6,*)' '
 1000 FORMAT (A4)
    1 CONTINUE
      READ (5,1000) REPLY
C
      IF (REPLY .EQ. ''R'') THEN
         WRITE (6,*)' R IS USED TO ADD A RESISTOR TO THE NETWORK.'
         WRITE (6,*)' '
         WRITE (6,*)' Type R <CR>'
         WRITE (6,*)' The program prompts for nodes to which the'
         WRITE (6,*)' resistor is connected, and for R in Kohms.'
         WRITE (6,*)' '
      ENDIF
C
      IF (REPLY .EQ. ''E'') THEN
```

```
         WRITE (6,*)' E IS USED TO ADD A VOLTAGE SOURCE TO THE NETWORK.'
         WRITE (6,*)' '
         WRITE (6,*)' Type E <CR>'
         WRITE (6,*)' The program prompts for:'
         WRITE (6,*)'      * the nodes to which the source is connected'
         WRITE (6,*)'      * the value of the source, in volts'
         WRITE (6,*)'      * the internal resistance, in Kohms'
         WRITE (6,*)' '
      ENDIF
C
      IF (REPLY .EQ. ''I'') THEN
         WRITE (6,*)' I IS USED TO ADD A CURRENT SOURCE TO THE NETWORK.'
         WRITE (6,*)' '
         WRITE (6,*)' Type I <CR>'
         WRITE (6,*)' The program prompts for:'
         WRITE (6,*)'      * the nodes to which the source is connected'
         WRITE (6,*)'      * the value of the source, in amps'
         WRITE (6,*)' '
      ENDIF
C
      IF (REPLY .EQ. ''LIST'') THEN
         WRITE (6,*)' LIST PRODUCES A LISTING OF THE NETWORK DATA.'
         WRITE (6,*)' '
         WRITE (6,*)' Type LIST <CR>'
      ENDIF
C
      IF (REPLY .EQ. ''VTVM'') THEN
         WRITE (6,*)' VTVM PRODUCES A LISTING OF NODE VOLTAGES.'
         WRITE (6,*)' '
         WRITE (6,*)' Type VTVM <CR>'
         WRITE (6,*)' VTVM will compute node voltages and'
         WRITE (6,*)'      print selected node voltages.'
         WRITE (6,*)' NODES is used to control which node voltages'
         WRITE (6,*)'      are printed.'
      ENDIF
C
      IF (REPLY .EQ. ''NODE'') THEN
         WRITE (6,*)' NODES SETS THE NODE PAIR VOLTAGES THAT'
         WRITE (6,*)' ARE COMPUTED AND PRESENTED.'
         WRITE (6,*)' '
         WRITE (6,*)' Type NODES <CR>'
         WRITE (6,*)' NODES will prompt for the number of node pairs'
         WRITE (6,*)' and the positive and negative node in each pair.'
      ENDIF
C
      IF (REPLY .EQ. ''MDFY'') THEN
         WRITE(6,*)' MDFY WILL CHANGE COMPONENT TYPES, VALUES'
         WRITE(6,*)' OR NODE CONNECTIONS.'
         WRITE(6,*)' USE ELIM TO REMOVE AN ELEMENT COMPLETELY.'
         WRITE (6,*)' '
         WRITE (6,*)' Type MDFY <CR>'
         WRITE (6,*)' MDFY will prompt you for:'
         WRITE (6,*)'    1. The element to change'
         WRITE (6,*)'    2. Change in element type'
         WRITE (6,*)'    3. Change in element value'
         WRITE (6,*)'    4. Changes in node connections'
      ENDIF
C
      IF (REPLY .EQ. ''ELIM'') THEN
         WRITE (6,*)' ELIM ELIMINATES AN ELEMENT FROM THE ELEMENT LIST.'
         WRITE (6,*)' '
         WRITE (6,*)' Type ELIM <CR>'
         WRITE (6,*)' ELIM will prompt for the number of the'
         WRITE (6,*)' element to be eliminated.'
      ENDIF
```

```
C
      IF (REPLY .EQ. ''UNIT'') THEN
         WRITE (6,*)' Units used in this program are'
         WRITE (6,*)' Volts, Amps and Kohms.'
         WRITE (6,*)' '
         WRITE (6,*)' Be careful. All resistor inputs are in Kohms!'
      ENDIF
C
      IF (REPLY .EQ. ''SAVE'') THEN
         WRITE (6,*)' SAVE permits you to save a circuit description'
         WRITE (6,*)' in a file for retrieval later.'
         WRITE (6,*)' The file name must be four characters or less.'
         WRITE (6,*)' SAVE will prompt you for a file name.'
      ENDIF
C
      IF (REPLY .EQ. ''GET'') THEN
         WRITE (6,*)' GET permits retrieval of a circuit description.'
         WRITE (6,*)' stored previously with a SAVE command.'
         WRITE (6,*)' GET will prompt you for the file name.'
      ENDIF
C
      RETURN
      END
```

Program Segments for Controlled Sources in BU-CAP

This appendix contains a program fragment for updating a conductance matrix and source vector when controlled sources are present. The symbols used are

- *J*   Current-controlled current source
- *K*   Voltage-controlled current source
- *U*   Current-controlled voltage source
- *V*   Voltage-controlled voltage source

```
C
      IF (KINDS(I) .EQ. ''J'') THEN
         NP = NODES(I,1)
         NM = NODES(I,2)
         NCONTROL = NODES(I,3)
         GI = VALUES(I)
         NCP = NODES(NCONTROL,1)
         NCM = NODES(NCONTROL,2)
         IF(KINDS(NCONTROL).EQ.''R'') GADD = GI/VALUES(NCONTROL)
         IF(KINDS(NCONTROL).EQ.''E'') GADD = GI/SUPP(NCONTROL)
         IF(KINDS(NCONTROL).EQ.''I'') GADD = 0.
C
         IF(KINDS(NCONTROL).EQ.''E'') THEN
            SADD = GI*VALUES(NCONTROL)/SUPP(NCONTROL)
         ENDIF
         IF(KINDS(NCONTROL).EQ.''I'') SADD = GI*VALUES(NCONTROL)
         IF(KINDS(NCONTROL).EQ.''R'') SADD = 0.
C
      ENDIF
C
      IF (KINDS(I) .EQ. ''V'') THEN
         NP = NODES(I,1)
         NM = NODES(I,2)
         NCP = NODES(I,3)
         NCM = NODES(I,4)
         GADD = VALUES(I)
```

```
            GADD = GADD + (1./SUPP(I))
            SADD = 0.
         ENDIF
C

         IF (KINDS(I) .EQ. ''U'') THEN
            NP = NODES(I,1)
            NM = NODES(I,2)
            NCONTROL = NODES(I,3)
            GI = VALUES(I)
            NCP = NODES(NCONTROL,1)
            NCM = NODES(NCONTROL,2)
            IF(KINDS(NCONTROL).EQ.''R'') GADD = GI/VALUES(NCONTROL)
            IF(KINDS(NCONTROL).EQ.''E'') GADD = GI/SUPP(NCONTROL)
            IF(KINDS(NCONTROL).EQ.''I'') GADD = 0.
            GADD = GADD + (1./SUPP(I))
C

            IF(KINDS(NCONTROL).EQ.''E'') THEN
                SADD = GI*VALUES(NCONTROL)/SUPP(NCONTROL)
            ENDIF
            IF(KINDS(NCONTROL).EQ.''I'') SADD = GI*VALUES(NCONTROL)
            IF(KINDS(NCONTROL).EQ.''R'') SADD = 0.
C

         ENDIF
         IF((KINDS(I).EQ.''U'').OR.(KINDS(I).EQ.''V'')
      1  .OR.(KINDS(I).EQ.''J'').OR.(KINDS(I).EQ.''K'')) THEN
C

            IF((NP.GT.0).AND.(NCP.GT.0)) THEN
               G(NP,NCP) = G(NP,NCP) + GADD
            ENDIF
C

            IF((NM.GT.0).AND.(NCP.GT.0)) THEN
               G(NM,NCP) = G(NM,NCP) - GADD
            ENDIF
C

            IF((NP.GT.0).AND.(NCM.GT.0)) THEN
               G(NP,NCM) = G(NP,NCM) - GADD
            ENDIF
C

            IF((NM.GT.0).AND.(NCM.GT.0)) THEN
               G(NM,NCM) = G(NM,NCM) + GADD
            ENDIF
C

            IF(NP.GT.0) SOURCE(NP) = SOURCE(NP) + SADD
            IF(NM.GT.0) SOURCE(NM) = SOURCE(NM) - SADD
C

         ENDIF
```

Some Programs for Examples in Chapter 2: Nonlinear Analysis

This appendix contains the source code for the example calculations that are given in Tables 2.1, 2.2, and 2.3.

```
C
C PROGRAM FOR SOLVING EXAMPLE
C NETWORK USING CONTRACTION MAPPING.
C
      WRITE (6,*)' TYPE SOURCE VOLTAGE AND INITIAL GUESS FOR V.'
      READ (5,*) VSOURCE, V
      WRITE (6,*)' TYPE THE NUMBER OF ITERATIONS.'
      READ (5,*) N
      WRITE (6,*)' TYPE THE VALUE OF THE RESISTOR (OHMS).'
      READ (5,*) R
      WRITE (6,*)' ITERATION    VOLTAGE'
      WRITE (6,*)' '
      DO 100 I = 1,N
      V = (-R*.1*(V**.6)) + VSOURCE
      WRITE (6,*) I,V
  100 CONTINUE
      CALL EXIT
      END
```

```
C
C PROGRAM FOR SOLVING THE SIMPLE NONLINEAR
C NETWORK USING NEWTON'S ALGORITHM.
C
      WRITE (6,*)' TYPE SOURCE VOLTAGE AND INITIAL GUESS FOR V.'
      READ (5,*) VSOURCE, V
      WRITE (6,*)' TYPE THE NUMBER OF ITERATIONS.'
      READ (5,*) N
      WRITE (6,*)' TYPE THE VALUE OF THE RESISTOR (OHMS).'
      READ (5,*) R
      WRITE (6,*)' ITERATION    VOLTAGE'
      WRITE (6,*)' '
```

```
      DO 100 I = 1,N
      G = (.1)*(V**.6)
      GPRIME = (.06)*(V**(-.4))
      F = G + (V-VSOURCE)/R
      FPRIME = GPRIME + 1./R
      V = V - (F/FPRIME)
      WRITE (6,*) I,V
100 CONTINUE
      CALL EXIT
      END


C
C PROGRAM FOR SOLVING THE RESISTOR-DIODE
C NETWORK USING NEWTON'S ALGORITHM.
C
      REAL I0
      WRITE (6,*)' TYPE SOURCE VOLTAGE AND INITIAL GUESS FOR V.'
      READ (5,*) VSOURCE, V
      WRITE (6,*)' TYPE THE NUMBER OF ITERATIONS.'
      READ (5,*) N
      WRITE (6,*)' TYPE THE VALUE OF THE RESISTOR (OHMS).'
      READ (5,*) R
      I0 = 1.E-9
      QKT = 40.
      WRITE (6,*)' ITERATION     VOLTAGE'
      WRITE (6,*)' '
      DO 100 I = 1,N
      G = I0*(EXP(QKT*V)-1.)
      GPRIME = (I0*QKT)*EXP(QKT*V)
      F = G + (V-VSOURCE)/R
      FPRIME = GPRIME + 1./R
      V = V - (F/FPRIME)
      WRITE (6,*) I,V
100 CONTINUE
      CALL EXIT
      END
```

Programs for Nonlinear Analysis

This appendix contains the source code for routines that permit solution of circuits with "standard" diodes. This code can be embedded within the circuit analysis program, and called whenever the user wants a solution for a network containing one or more diodes (Type D).

```
C
C********************************************************************
C
      SUBROUTINE CIRCSOLVE (NELMT,NOUT,NODOUT,VSOLVE)
      COMMON/CIRCUIT/NODES(20,2),VALUES(20),SUPP(20)/CHAR/KINDS(20)
      COMMON/INFO/CIRCTYPE,SIMPROB,OPTPROB
      DIMENSION NODOUT(20,2)
      DIMENSION VSOLVE(20)
      CHARACTER*1 KINDS
      CHARACTER*4 CIRCTYPE,SIMPROB,OPTPROB
      IF (CIRCTYPE .EQ. ''LIN'') THEN
          CALL LINSOLVE(NELMT,NOUT,NODOUT,VSOLVE)
      ELSE
          CALL NONLINSOLVE(NELMT,NOUT,NODOUT,VSOLVE)
      ENDIF
C
      RETURN
      END
C
C********************************************************************
C
      SUBROUTINE NONLINSOLVE(NELMT,NOUT,NODOUT,VSOLVE)
      COMMON/CIRCUIT/NODES(20,2),VALUES(20),SUPP(20)/CHAR/KINDS(20)
      DIMENSION NODOUT(20,2), V(20)
      DIMENSION NODED(20,2), NELDIOD(20)
      DIMENSION VSOLVE(20)
      CHARACTER*1 KINDS
      REAL I0
C
C DEFINITION OF VARIABLES
C
```

```
C         IO = REVERSE SATURATION CURRENT FOR A DIODE
C         QKT IS A SINGLE SYMBOL FOR Q/KT
C
C*******************************************************************
          QKT = 40.
          IO = 1.E-9
          NDIODES=0
          ITERD = 15
          DO 800 J = 1,NELMT
             IF (KINDS(J) .EQ. ''D'') THEN
                KINDS(J) = ''E''
                NDIODES=DIODES+1
                NELDIOD(NDIODES)=J
                NODED(NDIODES,1) = NODES(J,1)
                NODED(NDIODES,2) = NODES(J,2)
                VALUES(J)=(1./QKT)*((1./EXP(QKT*.5))+(QKT*.5-1.))
                SUPP(J)= 1./(IO*QKT*(EXP(QKT*.5)))
             ENDIF
  800     CONTINUE
          CALL LINSOLVE(NELMT,NDIODES,NODED,V)
          DO 820 J = 1,ITERD
             DO 810 JJ = 1,NDIODES
                ND = NELDIOD(JJ)
                KINDS(ND) = ''E''
                IF (V(JJ).LE. -1.) V(JJ) = -1.
                IF (V(JJ).GT. .6) V(JJ) = .6
                VALUES(ND)=(1./QKT)*((1./EXP(QKT*V(JJ)))+(QKT*V(JJ)-1.)))
                SUPP(ND) = 1./(IO*QKT*(EXP(QKT*V(JJ))))
  810        CONTINUE
             CALL LINSOLVE(NELMT,NDIODES,NODED,V)
  820     CONTINUE
        CALL LINSOLVE(NELMT,NOUT,NODOUT,VSOLVE)
          DO 830 J = 1,NDIODES
             ND=NELDIOD(J)
             KINDS(ND) = ''D''
             VALUES(ND) = 0.
             SUPP(ND) = 0.
  830     CONTINUE
C
      RETURN
      END
C
C*******************************************************************
C
```

Program Segment to Incorporate Capacitor and Inductor Companion Models for Time Response Calculation

This appendix contains the code for generating companion models for capacitors and inductors and computation of transient response using those models.

```
C
C
C      ****************************************************************
C
C      SUBROUTINE TRAN(NELMT,NOUT,NODOUT)
C
C      ****************************************************************
C
C      SUBROUTINE TITLE:   TRAN (TRANSIENT ANALYSIS)
C      WRITTEN BY:         MARK SWARTZ
C      DATE LAST REVISED:  APRIL 25, 1984
C
C      ****************************************************************
C
       SUBROUTINE TRAN(NELMT,NOUT,NODOUT)
       COMMON/CIRCUIT/NODES(20,2),VALUES(20),SUPP(20)/CHAR/KINDS(20)
      ·COMMON/PLOT/ YMIN,YMAX,NCOUNT
       COMMON/INFO/CIRCTYPE,OUTTYPE,SIMPROB,OPTPROB
       CHARACTER*4 CIRCTYPE,OUTTYPE,SIMPROB,OPTPROB
       DIMENSION NODOUT(20,2),VSOLVE(20),
      1TEMPK(20),TEMPV(20),TEMPS(20),NODENST(20,2),LOENST(20)
       CHARACTER*1 KINDS,TEMPK,SYMBOL
C
       WRITE(6,*)' PLEASE INPUT A DT AND A TOTAL TIME FOR THE ANALYSIS.'
       READ(5,*)DT,TIME
       WRITE(6,*)' PLEASE INPUT A PRINT INTERVAL--MUST BE AN INTEGER'
       WRITE(6,*)' MULTIPLE OF DT.'
       READ(5,*)TINTER
       WRITE(6,*)' THANKS.'
C
       T=0.
       NENRST=0
       NCAPS=0
       NINDS=0
C
```

```
C LOOK FOR ALL CAPACITORS AND INDUCTORS - - WHEN ONE IS FOUND,
C STORE ELEMENT DATA IN TEMPORARY ARRAYS.
C
      DO 10 I=1,NELMT
         IF (KINDS(I) .EQ. ''C'') THEN
            NENRST = NENRST + 1
            TEMPK(I)=KINDS(I)
            TEMPK(I)=VALUES(I)
            TEMPS(I)=SUPP(I)
            KINDS(I)=''E''
            VALUES(I)=SUPP(I)
            SUPP(I)=DT/TEMPV(I)
            LOENST(NENRST)=I
            NODENST(NENRST,1)=NODES(I,1)
            NODENST(NENRST,2)=NODES(I,2)
         END IF
C
         IF (KINDS(I) .EQ. ''L'') THEN
            NENRST = NENRST + 1
            TEMPK(I)=KINDS(I)
            TEMPV(I)=VALUES(I)
            TEMPS(I)=SUPP(I)
            KINDS(I)=''E''
            VALUES(I)=SUPP(I)
            SUPP(I)=TEMPV(I)/DT
            LOENST(NENRST)=I
            NODENST(NENRST,1)=NODES(I,1)
            NODENST(NENRST,2)=NODES(I,2)
         END IF
10    CONTINUE
C
      NPRIN=TINTER/DT
      NTIME = (TIME/DT)/NPRIN
C
C
      DO 11 I = 1,NOUT
         VSOLVE(I) = 0.
   11 CONTINUE
C
      IF (OUTTYPE .EQ. ''PLOT'') THEN
         WRITE(6,*)' PLOT SYMBOLS:'
         WRITE(6,*)' '
         DO 720 I=1,NOUT
            IF(I .EQ. 1) SYMBOL = ''*''
            IF(I .EQ. 2) SYMBOL = ''$''
            IF(I .EQ. 3) SYMBOL = ''#''
            IF(I .EQ. 4) SYMBOL = ''@''
            IF(I .EQ. 5) SYMBOL = ''&''
C
            WRITE (6,*)'V',NODOUT(I,1),'-V',NODOUT(I,2),' = ',SYMBOL
  720       CONTINUE
            NCOUNT = -1
            CALL PLOTPT (T,VSOLVE,NOUT)
      ENDIF
C
      IF (OUTTYPE .EQ. ''PRNT'')CALL PRINTPT(NOUT,NODOUT,VSOLVE,T)
C
      DO 20 J=1,NTIME
         DO 17 K=1,NPRIN
            T=T+DT
            CALL CIRCSOLVE(NELMT,NENRST,NODENST,VSOLVE)
C
            DO 15 I=1,NENRST
               M=LOENST(I)
C
```

```
              IF (TEMPK(M) .EQ. ''C'') THEN
                 VALUES(M)=VSOLVE(I)
              END IF
C
              IF (TEMPK(M) .EQ. ''L'') THEN
                 VALUES(M)=VALUES(M) - VSOLVE(I)
              END IF
C
15         CONTINUE
C
           CALL CIRCSOLVE(NELMT,NOUT,NODOUT,VSOLVE)
C
17      CONTINUE
C
     IF (OUTTYPE .EQ. ''PLOT'')CALL PLOTPT (T,VSOLVE,NOUT)
     IF (OUTTYPE .EQ. ''PRNT'')CALL PRINTPT(NOUT,NODOUT,VSOLVE,T)
C
20   CONTINUE
C
     DO 30 I=1,NENRST
        M=LOENST(I)
        KINDS(M)=TEMPK(M)
        VALUES(M)=TEMPV(M)
        SUPP(M)=TEMPS(M)
30   CONTINUE
C
     RETURN
     END
C
C*************************************************************************
C
     SUBROUTINE PRINTPT(NOUT,NODOUT,VSOLVE,T)
     DIMENSION NODOUT(20,2),VSOLVE(20)
C
     IF (T .EQ. 0.) THEN
       WRITE (6,*)' NODE DIFFERENCES OUTPUT ARE:'
       DO 10 K = 1,NOUT
         WRITE(6,*)'V',NODOUT(K,1),'-V',NODOUT(K,2)
   10    CONTINUE
       WRITE (6,*) ' '
       WRITE (6,*)'TIME (SEC)     NODE VOLTAGE DIFFERENCES'
       WRITE (6,*) ' '
     ENDIF
C
     WRITE (6,1000) T,(VSOLVE(K), K=1,NOUT)
 1000 FORMAT (8E10.4)
C
     RETURN
     END
```

Demonstration Program for the Backward Euler Algorithm and Other Implicit Algorithms

This appendix contains a demo program that can be used to check out various implicit algorithms, starting with the backward Euler.

```
      DIMENSION X(10)
      CALL SIMIN (X,N,DT,NSTEPS)
      CALL SIMRUN (X,N,DT,NSTEPS)
      CALL EXIT
      END
C
C
C *******************************
C
C
      SUBROUTINE XDOT(X,XD,T,N)
      DIMENSION X(10), XD(10)
      XD(1) = -X(1)
      XD(2) = -5.*X(2)
      RETURN
      END
C
C****************************************************************
C
      SUBROUTINE SIMIN (X,N,DT,NSTEPS)
      DIMENSION X(20)
      WRITE (6,*)' INPUT DATA FOR THE SIMULATION.'
      WRITE (6,*)' INPUT THE NUMBER OF STATES (I5 FORMAT).'
      READ (5,*) N
      WRITE (6,*)' INPUT THE INITIAL STATES (E10.4 FORMAT).'
      READ (5,*) (X(J), J=1,N)
      WRITE (6,*)' INPUT THE INTEGRATION INTERVAL (E10.4 FORMAT).'
      READ (5,*) DT
      WRITE (6,*)' INPUT THE NUMBER OF STEPS IN THE RUN (I5 FORMAT).'
      READ (5,*) NSTEPS
      TIME = DT*NSTEPS
      WRITE (6,*)' THE TOTAL SIMULATION TIME (SECONDS) IS:'
 1005 FORMAT (1X,E15.6)
```

```
      WRITE (6,1005) TIME
      RETURN
      END
C
C****************************************************************
C
      SUBROUTINE SIMRUN (X,N,DT,NSTEPS)
      DIMENSION X(20)
      COMMON/IMPLICIT/DELX(20)
          WRITE (6,*)' INPUT INCREMENTS FOR JACOBIAN EVALUATION.'
          READ (5,*) (DELX(I), I=1,N)
      T = 0.
 1111 FORMAT (' TIME           STATES')
      PRINT 1111
 1001 FORMAT (1X,8(E12.6,4X))
      PRINT 1001, T, (X(I), I=1,N)
      DO 100 IT = 1,NSTEPS
      CALL BACKEULER (X,T,DT,N)
      PRINT 1001, T,(X(I), I=1,N)
  100 CONTINUE
      RETURN
      END
C
C****************************************************************
C
      SUBROUTINE BACKEULER (X,T,DT,N)
      DIMENSION X(20), XD(20), JACOB(20,20), AINV(20,20)
      DIMENSION DELX(20)
      REAL JACOB
C
C   BACKWARD EULER INTEGRATION ALGORITHM
C
      CALL JACOBIAN (X,JACOB,T,N)
      DO 20 I = 1,N
      DO 20 J = 1,N
          IF (I .EQ. J) THEN
              JACOB(I,J) = 1. - DT*JACOB(I,J)
          ELSE
              JACOB(I,J) = -DT*JACOB(I,J)
          ENDIF
   20 CONTINUE
      CALL MATIG (JACOB,AINV,N)
      CALL XDOT (X,XD,T,N)
      DO 100 I = 1,N
          DO 50 J = 1,N
              X(I) = X(I) + AINV(I,J)*DT*XD(J)
   50     CONTINUE
  100 CONTINUE
      T = T+DT
      RETURN
      END
C
C****************************************************************
C
      SUBROUTINE JACOBIAN (X,JACOB,T,N)
C
C JACOBIAN computes the Jacobian of the derivative vector
C F(X,U,t) computed by XDOT(X,XD,T,N).
C
      DIMENSION JACOB(20,20), X(20)
      DIMENSION XD(20), DELXD(20)
      COMMON/IMPLICIT/DELX(20)
      REAL JACOB
C
      CALL XDOT(X,XD,T,N)
```

```
          DO 100 J = 1,N
              X(J) = X(J) + DELX(J)
              CALL XDOT(X,DELXD,T,N)
                 DO 50 I = 1,N
                     JACOB(I,J) = (DELXD(I)-XD(I))/DELX(J)
   50            CONTINUE
              X(J) = X(J) - DELX(J)
  100   CONTINUE
        RETURN
        END
C
C*************************************************************
C

SECOND ORDER GEAR ALGORITHM AND TRAPEZOIDAL ALGORITHM

C
C*************************************************************
C
        SUBROUTINE GEAR2 (X,T,DT,N)
        DIMENSION X(20), XD(20), JACOB(20,20), AINV(20,20)
        DIMENSION XSAVE(20),XTEMP(20)
        SAVE XSAVE
        REAL JACOB
C
C SECOND ORDER GEAR INTEGRATION ALGORITHM
C
C
C INITIALIZE Xn-1
C
        IF (T .LT. 1.E-20) THEN
           DO 10 I = 1,N
              XSAVE(I) = X(I)
   10      CONTINUE
        ENDIF
C
        CALL JACOBIAN (X,JACOB,T,N)
        DO 20 I - 1,N
        DO 20 J = 1,N
           IF (I .EQ. J) THEN
                 JACOB(I,J) = 1. -(2.*DT/3.)*JACOB(I,J)
           ELSE
                 JACOB(I,J) = -(2.*DT/3.)*JACOB(I,J)
           ENDIF
   20   CONTINUE
        DO 30 I = 1,N
           XTEMP(I) = XSAVE(I)
           XSAVE(I) = X(I)
   30   CONTINUE
C
        CALL MATIG (JACOB,AINV,N)
        CALL XDOT (X,XD,T,N)
        DO 100 I = 1,N
                 XSAVE(J) = X(J)
           DO 50 J = 1,N
                 XD(J) = (2./3.)*XD(J) +(X(J)-XTEMP(J))/(3.*DT)
                 X(I) = X(I) + AINV(I,J)*DT*XD(J)
   50      CONTINUE
  100   CONTINUE
        T = T+DT
        RETURN
        END
```

```
C
C***************************************************************
C
      SUBROUTINE MODTRAPEZ (X,T,DT,N)
      DIMENSION X(20), XD(20), JACOB(20,20), AINV(20,20)
      DIMENSION DELX(20)
      REAL JACOB
C
C  BACKWARD EULER INTEGRATION ALGORITHM
C
      CALL JACOBIAN (X,JACOB,T,N)
      DO 20 I = 1,N
      DO 20 J = 1,N
          IF (I .EQ. J) THEN
              JACOB(I,J) = 1. - (DT/2.)*JACOB(I,J)
          ELSE
              JACOB(I,J) = -(DT/2.)*JACOB(I,J)
          ENDIF
   20 CONTINUE
      CALL MATIG (JACOB,AINV,N)
      CALL XDOT (X,XD,T,N)
      DO 100 I = 1,N
          DO 50 J = 1,N
              X(I) = X(I) + AINV(I,J)*DT*XD(J)
   50     CONTINUE
  100 CONTINUE
      T = T+DT
      RETURN
      END
```

Programs for Optimization Problems (Chapters 5 and 6)

```
      SUBROUTINE GOLSEC (XLEFT,XRIGHT,ITER)
C
C GOLSEC DOES A GOLDEN SECTION SEARCH
C
      X1 = XLEFT + (.381966)*(XRIGHT-XLEFT)
      X2 = XLEFT + (.618034)*(XRIGHT-XLEFT)
      CALL F(X1,F1)
      CALL F(X2,F2)
      DO 100 I=1,ITER
         IF (F1 .GT. F2) THEN
            XLEFT = X1
            FLEFT = F1
            X1 = X2
            F1 = F2
            X2 = XLEFT + (.618034)*(XRIGHT-XLEFT)
            CALL F(X2,F2)
         ELSE
            XRIGHT = X2
            FRIGHT = F2
            X2 = X1
            F2 = F1
            X1 = XLEFT + (.381966)*(XRIGHT-XLEFT)
            CALL F(X1,F1)
         ENDIF
  100 CONTINUE
      RETURN
      END
C
C
C
      SUBROUTINE FOFX(X,N,F)
      DIMENSION X(10)
C
C THIS FUNCTION IS THE ROSENBROCK FUNCTION.
C
      F = 100.*(X(1)**2 - X(2))**2 + (1.-X(1))**2
      RETURN
      END
```

```
C
C
C
C
C************************************************
C
      SUBROUTINE SIMPLEX (X,N)
C
C
C SIMPLEX IMPLEMENTS THE SIMPLEX SEARCH ALGORITHM
C FOR MINIMIZING FOFX.
C
C
      DIMENSION X(10), XTEST(10), XSIMPLEX(10,11)
      DIMENSION XNEW(10), F(11), INDEX(11)
      COMMON/OPT/SCALE(10)
C
      P = (SQRT(N+1.)+N-1.)/(N*SQRT(2.))
      Q = (SQRT(N+1.)-1.)/(N*SQRT(2.))
C
      DO 900 NNN = 1,4
      DO 20 J = 1,N
      DO 19 I = 1,N
         IF (I .EQ. J) THEN
            XSIMPLEX(J,I) = X(J) + P*SCALE(J)
         ELSE
            XSIMPLEX(J,I) = X(J) + Q*SCALE(J)
         ENDIF
   19 CONTINUE
   20 CONTINUE
      DO 25 J = 1,N
         XSIMPLEX(J,N+1) = X(J)
   25 CONTINUE
C

C
      DO 40 K = 1,N+1
         DO 30 J = 1,N
            XTEST(J) = XSIMPLEX(J,K)
   30    CONTINUE
         CALL FOFX (XTEST,N,F(K))
   40 CONTINUE
C
C Sort function values
C
      DO 45 L = 1,N+1
         INDEX(L) = L
   45 CONTINUE
C
      DO 600 ITER = 1,30
C
      DO 50 L = 1,N
      DO 50 K = 1,N
         IF (F(K+1) .LT. F(K)) THEN
            FTEMP = F(K)
            F(K) = F(K+1)
            F(K+1) = FTEMP
            ITEMP = INDEX(K)
            INDEX(K) = INDEX(K+1)
            INDEX(K+1) = ITEMP
         ENDIF
   50 CONTINUE
C
C Function Values Now Are Sorted, Max in INDEX(N+1)
C
```

```
             IF (ITER .EQ. 1) THEN
                IMAXLAST = INDEX(N+1)
                IMINLAST = INDEX(1)
                IMINCOUNT = 1
                IREFLECT = INDEX(N+1)
             ELSE
                IF (INDEX(1) .EQ. IMINLAST) THEN
                   IMINCOUNT=IMINCOUNT+1
                ELSE
                   IMINLAST = INDEX(1)
                   IMINCOUNT = 1
                ENDIF
                IF (IMINCOUNT .GT. 2*N) GO TO 800
                IF (INDEX(N+1) .EQ. IMAXLAST) THEN
                   IREFLECT=INDEX(N)
                   IMAXLAST=IREFLECT
                ELSE
                   IREFLECT=INDEX(N+1)
                   IMAXLAST=IREFLECT
                ENDIF
             ENDIF
  140 CONTINUE
C
      DO 200 J = 1,N
         XNEW(J) = 0.
         DO 150 I = 1,N+1
            XNEW(J) = (2./N)*XSIMPLEX(J,I) + XNEW(J)
  150    CONTINUE
         XNEW(J) = XNEW(J) - XSIMPLEX(J,IREFLECT)*(1.+(2./N))
  200 CONTINUE
C
      DO 300 J = 1,N
         XSIMPLEX(J,IREFLECT) = XNEW(J)
  300 CONTINUE
C
      DO 310 J = 1,N+1
         IF (INDEX(J) .EQ. IREFLECT) IMAX = J
  310 CONTINUE
      CALL FOFX(XNEW,N,F(IMAX))
C
  600 CONTINUE
  800 CONTINUE
      DO 890 J = 1,N
         X(J) = XSIMPLEX(J,IMINLAST)
  890 CONTINUE
      P = .1*P
      Q = .1*Q
  900 CONTINUE
      RETURN
      END
C
C This is a mainline and subroutines for use in applying
C the Fletcher-Powell algorithm for function minimization.
C
      DIMENSION GRD(10), S(10), X(10), DX(10)
      DIMENSION G(10,10)
C
C*****INITIALIZE VARIABLES******************
C
      DO 101   I=1,10
         GRD(I) = 0.
         S(I) = 0.
         X(I) = 0.
         DX(I) = 0.
  101 CONTINUE
      N = 1
```

```
C
C*****READ IN THE INITIAL GUESS FOR THE OPTIMUM LOCATION.******
C
      WRITE (6,*) ' INPUT THE NUMBER OF PARAMETERS.'
      READ (5,*) N
      WRITE (6,*) ' INPUT THE STARTING VALUES OF PARAMETERS.'
      READ (5,*) (X(I), I=1,N)
      WRITE (6,*) ' INPUT THE STEP INCREMENT TO BE USED WHEN'
      WRITE (6,*) ' FINDING NEW X AND DX VALUES.'
      READ (5,*) STEPSZ
      WRITE (6,*) ' HOW MANY STEPS DO YOU WANT TO DO'
      WRITE (6,*) ' IN YOUR OPTIMIZATION RUN?'
      READ (5,*) NSTEPS

      WRITE (6,*) ' PARAMETER VALUES AND FUNCTION VALUE'
C
C
C*********INITIALIZE G MATRIX, GRADIENT, AND F(X) VALUES*********
C
      CALL MATIDN (G,N)
      CALL GRAD (X,GRD,N)
      CALL FOFX (X,N,FX)
C
C*********PRINT AND SAVE INITIAL VALUES*****************
C
      WRITE(6,*) ' BEGINNING VALUES BEFORE OPTIMIZATION'
      WRITE(6,*) 'F(X) = ',FX
      DO 111 I = 1,N
          WRITE(6,*) 'X',I, '=', X(I), 'DX',I, '=', DX(I)
  111 CONTINUE

      DO 100 ITER=1,NSTEPS
C*********FIND S - DIRECTION VECTOR***********
C
          DO 300 I=1,N
              S(I)=0.
              DO 400 J=1,N
                  S(I) = S(I) - G(I,J) * GRD(J)
  400         CONTINUE
  300     CONTINUE
C
C*********MOVE DX AND LET X(NEW) = X(OLD) + DX*********
C
          CALL DIROPT(X,S,N,STEPSZ,DX)
C
C*********PRINT AND STORE ITERATION VALUES*************
C
          CALL FOFX(X,N,FX)
          WRITE(6,*) 'OPTIMIZATION ITERATION #', ITER
          WRITE(6,*) 'F(X) = ',FX
          DO 112 K=1,N
              WRITE(6,*) 'X',K, '= ', X(K), 'DX',K, '= ', DX(K)
  112     CONTINUE
C
C ************FIND THE NEW G MATRIX******************
C
  100 CONTINUE
      PRINT*,'FLETCHER POWELL OPTIMIZATION RUN IS COMPLETED'
      END


C
C*************F(X) CALCULATION SUBROUTINE************
C
C     This subroutine contains the function to be minimized by
C the Fletcher-Powell routine.
```

```
C
      SUBROUTINE FOFX(X,N,FX)

      DIMENSION X(10)

      FX = X(1)**2 + X(2)**2

      RETURN
      END
C
C     Subroutine GNEW will be used to adjust the G matrix so that
C it will be an increasingly better approximation of the inverse
C Hessian. The equation to find this is:
C
C                    T              T
C     G = G + (DX)*(DX)  -  (GY)*(GY)
C             ---------    ---------
C                 D1           D2
C
C                  T            T
C     Where D1 = Y (DX) and D2 = Y (GY), and
C
C     Y = GRAD(Xnew) - GRAD(Xold)
C
C     To make things more readable, GY = G*Y
C
C
C
      SUBROUTINE GNEW(G,X,DX,GRD,N)
C
      REAL X(10), DX(10), NGRD(10), GRD(10), Y(10), GY(10)
      REAL G(10,10)
C
C********INITIALIZE VARIABLES***********
C
      DO 107 I = 1,N
           GY(I) = 0.
            Y(I) = 0.
  107 CONTINUE

      D1 = 0.
      D2 = 0.
C
C*******FIND NEW GRADIENT (NGRAD) = GRAD(Xnew)************
C
      CALL GRAD(X,NGRD,N)

C*********FIND Y***************
C
      DO 100 I=1,N
           Y(I) = NGRD(I) - GRD(I)
           GRD(I) = NGRD(I)
  100 CONTINUE
C
C***********FIND GY*************
C
      CALL MATSCALE(G,Y,GY,N)
C
C*********FIND D1*************
C
      DO 200 I=1,N
           D1=D1 + Y(I) * DX(I)
  200 CONTINUE
```

```
C
C*********FIND D2*************
C
      DO 300 I=1,N
          D2=D2 + GY(I) * Y(I)
  300 CONTINUE
C
C*********FIND THE NEW G MATRIX VALUES************
C
      DO 400 I=1,N
          DO 500 J=1,N
              G(I,J) = G(I,J) + DX(I)*DX(J)/D1 - GY(I)*GY(J)/D2
  500     CONTINUE
  400 CONTINUE
      RETURN
      END


C*************************************************
C
      SUBROUTINE GRAD (X,G,N)
C
C GRAD EVALUATES THE GRADIENT OF A FUNCTION, FOFX
C
C    X = POINT AT WHICH THE GRADIENT IS EVALUATED
C        X IS THE PARAMETER VECTOR.
C
C    STEP = INCREMENTS IN X-COMPONENTS
C           WHEN EVALUATING THE GRADIENT.** CURRENTLY INT GEN**
C
C    G = COMPUTED VALUE OF THE GRADIENT.
C
C    N = NUMBER OF COMPONENTS IN X.
C
      DIMENSION X(10), G(10), STEP(10)
      CALL FOFX (X,N,F)
      DO 100 I = 1,N
         STEP(I) = X(I)*.000001
         IF (ABS(X(I)) .LT. 1.E-12) STEP(I) = .000001
         X(I) = X(I) + STEP(I)
         CALL FOFX (X,N,FDEL)
         G(I) = (FDEL - F)/STEP(I)
         X(I) = X(I) - STEP(I)
  100 CONTINUE
      RETURN
      END
C
      SUBROUTINE DIROPT (X,D,N,A,DX)
      DIMENSION X(10), D(10), DX(10), XOLD(10)
C
C    X = PARAMETER VECTOR
C    D = Vector defining search direction (gradient)
C    DX= Vector difference between old parameter vector
C        and new parameter vector.
C    OLDX = Old parameter vector.
C
      ALPHA = A
C
C*******SAVE X(old) vector and find initial F(X) values*********
C
      DO 5 I = 1,N
         XOLD(I) = X(I)
    5 CONTINUE
      CALL FOFX (X,N,FOLD)
```

```
C
C***********Find X(new)*************
C
      DO 100 IK = 1,20
          DO 10 I = 1,N
               X(I) = X(I) + ALPHA*D(I)
   10     CONTINUE
C
C*****Find new F(X) value and compare with old F(X) value********
C     If F(X)new > F(X)old reverse direction and cut distance scale.
.C     If F(X)new < F(X)old double distance multiplier and continue.
C
          CALL FOFX (X,N,FNEW)
          IF (FNEW .GT. FOLD) THEN
              ALPHA = -ALPHA/5.
          ELSE IF (FNEW .LT. FOLD) THEN
              ALPHA = 2.*ALPHA
              FOLD = FNEW
          END IF
  100 CONTINUE
C
C******Find DX = X(new) - X(old)*************
C
      DO 109 I = 1,N
          DX(I) = X(I) - XOLD(I)
  109 CONTINUE
C
C******Save X(new) values***********
C
      RETURN
      END
C
C Subroutine MATIDN is used to initiate the G matrix which
C will be used to approximate the inverse Hessian matrix.
C This subroutine initiates the G matrix as the identity matrix.
C
      SUBROUTINE MATIDN(G,N)

      DIMENSION G(10,10)

      DO 100 I = 1,N
          DO 200 J = 1,N
              IF ( I .EQ. J ) THEN
              G(I,J) = 1.
              ELSE
              G(I,J) = 0.
              ENDIF
  200     CONTINUE
  100 CONTINUE

      RETURN
      END
C     The purpose of this subroutine is to multiply an NxN matrix
C     by a vector of maximum size N, giving another vector of size N
C
      SUBROUTINE MATSCALE(X1,X2,X3,N)
      DIMENSION X1(10,10), X2(10), X3(10)

      DO 100 I=1,N
          X3(I)=0.
          DO 200 J=1,N
              X3(I) = X3(I) + X1(I,J) * X2(J)
  200     CONTINUE
```

```
100   CONTINUE
      RETURN
      END

C     THE PURPOSE OF THIS SUBROUTINE IS TO COPY ONE VECTOR ARRAY
C ONTO ANOTHER VECTOR ARRAY FOR 10 OR LESS VALUES.
C
      SUBROUTINE VECCOP(X1,X2,N)

      DIMENSION X1(10), X2(10)

      DO 100 I=1,N
          X2(I) = X1(I)
100   CONTINUE
      RETURN
      END
```

Programs Manipulating Topological Circuit Descriptions

This appendix contains basic topological routines that will

* Form the incidence matrix (INCIDENCEMATRIX) from the arrays used in the material in Chapters 1 through 3. INCIDENCEMATRIX adds nodes for the internal nodes in voltage sources.

* Sort elements by type and value (CKTSRT) preparatory to forming a proper tree.

* Rearrange elements to locate a (proper) tree.

```
C
C******************************************************************
C
      SUBROUTINE INCIDENCEMATRIX(NELMT)
C
      COMMON/CIRCUIT/NODES(20,4),VALUES(20),SUPP(20)/CHAR/KINDS(20)
      CHARACTER*1 KINDS
      CHARACTER*4 CIRCTYPE,NODECOUNT
      COMMON/TOPO/INCID(19,20),ELVAL(20),KIND(20),NEL(20),NELMNT,NNODES
C
C******************************************************************
C
C        INCID CONTAINS THE REDUCED INCIDENCE MATRIX
C        ELVAL CONTAINS ELEMENT VALUES
C        KIND CONTAINS A NUMERICAL CODE FOR ELEMENT KINDS
C            (KINDS CONTAINS A LITERAL CODE-TAKE NOTE)
C        NEL CONTAINS THE ORDER OF THE ELEMENTS AFTER SORTING
C
C******************************************************************
C
C        INCIDENCEMATRIX FORMS THE INCIDENCE MATRIX FOR
C        TOPOLOGICAL ANALYSIS
C
C        NELMNT is set to a positive number, and
```

```
C      NELMNT>0 is used to test if the Incidence Matrix
C      has been formed (by other routines).
C
       CHARACTER*1 YESORNO
C
       NELMNT = NELMT
       DO 50 I = 1,19
       DO 45 J = 1,20
       INCID(I,J) = 0
   45 CONTINUE
   50 CONTINUE
       CALL NODEFINDER (NNODES,NODECOUNT,NELMT)
       IF(NODECOUNT .EQ. ''MISS'') THEN
          WRITE(6,*)'' NODE MISSING IN LIST. CHECK YOUR CIRCUIT.''
          RETURN
       ENDIF
C
C
C SET UP CIRCUIT DATA IN DIFFERENT MATRICES
C
       NODESADDED = 0
       CIRCTYPE = ''LIN''
       DO 200 I = 1,NELMT
          ELVAL(I) = VALUES(I)
          IF(KINDS(I) .EQ. ''E'') THEN
             KIND(I) = 1
             NNODES = NNODES + 1
             NODESADDED = NODESADDED + 1
             KIND(NELMT + NODESADDED) = 3
             ELVAL(NELMT + NODESADDED) = SUPP(I)
             NODES(NELMT + NODESADDED,1) = NODES(I,1)
             NODES(NELMT + NODESADDED,2) = NNODES
             NODES(I,1) = NNODES
          ENDIF
          IF(KINDS(I) .EQ. ''C'') KIND(I) = 2
          IF(KINDS(I) .EQ. ''R'') KIND(I) = 3
          IF(KINDS(I) .EQ. ''L'') KIND(I) = 4
          IF(KINDS(I) .EQ. ''I'') KIND(I) = 5
          IF(KINDS(I) .EQ. ''D'') CIRCTYPE = ''NONL''
  200 CONTINUE
       NELMNT = NELMNT + NODESADDED
       DO 210 I = 1,NELMNT
          NEL(I) = I
  210 CONTINUE
C
       IF (CIRCTYPE .EQ. ''NONL'') THEN
          WRITE (6,*)' ATTEMPTED TO GET STATE EQUATIONS FOR A NONLINEAR'
          WRITE (6,*)' CIRCUIT!   STEP ABORTED.'
          RETURN
       ENDIF
C
C FORM REDUCED INCIDENCE MATRIX FROM TOPOLOGICAL DATA
C
       DO 100 I = 1,NELMNT
          IF(NODES(I,1) .NE. 0) INCID(NODES(I,1),I) = +1
          IF(NODES(I,2) .NE. 0) INCID(NODES(I,2),I) = -1
  100 CONTINUE
C
C
       RETURN
       END
C
C****************************************************************
C
       SUBROUTINE CKTSRT
```

```
C
        COMMON/TOPO/INCID(19,20),ELVAL(20),KIND(20),NEL(20),NELMNT,NNODES
C
C
C CKTSRT SORTS CIRCUIT INFORMATION
C ELEMENTS ARE SORTED BY ELEMENT TYPE
C WITHIN TYPES, ELEMENTS ARE SORTED BY ELEMENT VALUE
C ARRAY NEL IS USED TO KEEP TRACK OF ELEMENT ORDER.
C
        DIMENSION N(5)
        N1=NELMNT-1
        DO 30 J=1,N1
        N2=NELMNT-J
        DO 25 I = 1,N2
        IF(KIND(I+1)-KIND(I)) 10,25,25
     10 ITEMP=KIND(I)
        KIND(I)=KIND(I+1)
        KIND(I+1)=ITEMP
        ITEMP=NEL(I)
        NEL(I)=NEL(I+1)
        NEL(I+1)=ITEMP
        TEMP=ELVAL(I)
        ELVAL(I)=ELVAL(I+1)
        ELVAL(I+1)=TEMP
        DO 20 L = 1,NNODES
        ITEMP=INCID(L,I)
        INCID(L,I)=INCID(L,I+1)
     20 INCID(L,I+1)=ITEMP
     25 CONTINUE
     30 CONTINUE
C
C ELEMENTS NOW ARE SORTED BY TYPE
C LOWEST TYPES ARE IN THE LEFT PART OF THE INCIDENCE MATRIX
C NOW DETERMINE THE NUMBER OF EACH ELEMENT TYPE
C N(I)=NUMBER OF ITH ELEMENT TYPE
C
        DO 40 I=1,5
     40 N(I)=0
C
C SORT WITHIN ELEMENT TYPES.
C
        DO 50 I=1,NELMNT
        L=KIND(I)
C
C REPLACE INDUCTANCE VALUES BY RECIPROCALS DURING SORTING.
C THIS LEAVES THE SMALLEST INDUCTORS IN THE TREE WHEN CKTREE IS USED.
C THE ORIGINAL INDUCTANCE VALUES ARE REPLACED AFTER THE TREE IS
C GENERATED IN CKTREE.
C COUNT NUMBER OF EACH TYPE OF ELEMENT
C STORE RESULT IN ARRAY N.
C
        IF (L-4) 49,41,49
     41 ELVAL(I) = 1./ELVAL(I)
     49 CONTINUE
     50 N(L)=N(L)+1
        N1=0
        N2=1
        DO 100 I=1,5
        N1=N1+N(I)
        N3=N(I)-1
        IF(N3) 90,90,55
     55 CONTINUE
        DO 80 J=1,N3
        N4=N1-J
        DO 80 K=N2,N4
```

```
        IF(ELVAL(K+1)-ELVAL(K)) 80,80,60
   60 ITEMP=NEL(K)
      NEL(K)=NEL(K+1)
      NEL(K+1)=ITEMP
      TEMP=ELVAL(K)
      ELVAL(K)=ELVAL(K+1)
      ELVAL(K+1)=TEMP
      DO 70 L=1,NNODES
      ITEMP=INCID(L,K)
      INCID(L,K)=INCID(L,K+1)
   70 INCID(L,K+1)=ITEMP
   80 CONTINUE
   90 CONTINUE
      N2=N2+N(I)
  100 CONTINUE
      RETURN
      END
C
C*****************************************************************
C
      SUBROUTINE CKTREE
      COMMON/TOPO/INCID(19,20),ELVAL(20),KIND(20),NEL(20),NELMNT,NNODES
C
C CKTREE FINDS THE PROPER TREE OF A CIRCUIT
C CKTREE USES THE REDUCED INCIDENCE MATRIX, INCID
C
      DO 100 I=1,NNODES
      I1 = I + 1
      NSWAP = I
C
C CHECK FOR NONZERO DIAGONAL ELEMENT
C CHANGE SIGNS IN ROW IF A NEGATIVE DIAGONAL ELEMENT IS FOUND.
C
      NTEST = 0
      IF (INCID(I,I) .EQ. -1) THEN
         NTEST = 1
         DO 15 J=1,NELMNT
            INCID(I,J)=-INCID(I,J)
   15    CONTINUE
         ENDIF
C
         IF (INCID(I,I) .EQ. 1) NTEST = 1
C
C NTEST WILL REMAIN ZERO IF NO NONZERO ELEMENT IS FOUND BELOW THE
C ZERO ELEMENT ON THE DIAGONAL
C CHECK FOR NONZERO ELEMENT BELOW THE DIAGONAL
C SWAP ROWS IF A NONZERO ELEMENT IS FOUND
C
      IF (NNODES .GE. I1) THEN
         DO 30 J=I1,NNODES
            IF(INCID(J,I) .LT. 0) THEN
               NTEST=1
               DO 27 K = I,NELMNT
                  ITEMP=INCID(I,K)
                  INCID(I,K)=-INCID(J,K)
   27             INCID(J,K)=ITEMP
            ENDIF
            IF(INCID(J,I) .GT. 0) THEN
               NTEST = 1
               DO 29 K=I,NELMNT
                  ITEMP=INCID(I,K)
                  INCID(I,K)=INCID(J,K)
   29             INCID(J,K)=-ITEMP
            ENDIF
   30    CONTINUE
```

```
      ENDIF
C
C IF NTEST=0, COLUMNS MUST BE INTERCHANGED IN THE INCIDENCE MATRIX
C SWAP PRESENT COLUMN AND COLUMN NSWAP
C IF NTEST=1, A NONZERO ELEMENT HAS BEEN FOUND AND
C THERE IS NOW A ONE ON THE MAIN DIAGONAL.
C
      IF(NTEST) 55,40,55
   40 CONTINUE
      NSWAP=NSWAP+1
C
C CHECK FOR A NONZERO ELEMENT IN COLUMN NSWAP
C SWAP IF A NONZERO ELEMENT IS FOUND
C
      DO 41 J=I,NNODES
      NONZ=J
      IF(INCID(J,NSWAP)) 42,41,42
   41 CONTINUE
      GO TO 43
C SET NTEST =1 IF A NONZERO ELEMENT IS FOUND
   42 NTEST=1
   43 CONTINUE
      IF(NTEST) 45,40,45
C
C SWAP COLUMNS
C
   45 ITEMP=NEL(I)
      NEL(I)=NEL(NSWAP)
      NEL(NSWAP)=ITEMP
      TEMP=ELVAL(I)
      ELVAL(I)=ELVAL(NSWAP)
      ELVAL(NSWAP)=TEMP
      ITEMP=KIND(I)
      KIND(I)=KIND(NSWAP)
      KIND(NSWAP)=ITEMP
      DO 50 J=1,NNODES
      ITEMP=INCID(J,I)
      INCID(J,I)=INCID(J,NSWAP)
      INCID(J,NSWAP)=ITEMP
   50 CONTINUE
C
C BRING NONZERO ELEMENT UP TO DIAGONAL
C
      IF (INCID(NONZ,I)) 51,58,53
   51 CONTINUE
      DO 52 J=1,NELMNT
      ITEMP=-INCID(NONZ,J)
      INCID(NONZ,J)=INCID(I,J)
   52 INCID(I,J)=ITEMP
      GO TO 55
   53 CONTINUE
      DO 54 J=1,NELMNT
      ITEMP=INCID(I,J)
      INCID(I,J)=INCID(NONZ,J)
   54 INCID(NONZ,J)=ITEMP
   55 CONTINUE
   58 CONTINUE
C
C ELIMINATE (ALA GAUSS) NONZERO TERMS IN THE ITH COLUMN
C ELIMINATE TERMS BELOW THE DIAGONAL
C RESULT WILL BE AN UPPER TRIANGULAR MATRIX.
C
      DO 70 J=I1,NNODES
      IF(INCID(J,I)) 60,70,65
   60 CONTINUE
```

```
          DO 61 K=I,NELMNT
   61  INCID(J,K)=INCID(J,K)+INCID(I,K)
          GO TO 70
   65  CONTINUE
          DO 66 K=I,NELMNT
   66  INCID(J,K)=INCID(J,K)-INCID(I,K)
   70  CONTINUE
  100  CONTINUE
C
C ELIMINATE UPPER TRIANGULAR TERMS TO PRODUCE IDENTITY MATRIX
C
          IF(NNODES-1) 200,200,105
  105  CONTINUE
          DO 150 I=2,NNODES
          I1=I-1
          DO 150 J=1,I1
          IF(INCID(J,I)) 110,150,120
  110  CONTINUE
          DO 115 K=I,NELMNT
  115  INCID(J,K)=INCID(J,K)+INCID(I,K)
          GO TO 150
  120  CONTINUE
          DO 125 K=I,NELMNT
  125  INCID(J,K)=INCID(J,K)-INCID(I,K)
  150  CONTINUE
C
C CONVERT LINK RESISTANCE VALUES TO CONDUCTANCE VALUES
C
          I1 = NNODES + 1
          DO 135 K = I1,NELMNT
          L = KIND(K)
          IF (L-3) 133,131,133
  131  CONTINUE
          ELVAL(K) = 1./ELVAL(K)
  133  CONTINUE
  135  CONTINUE
C
C CONVERT RECIPROCAL INDUCTANCE VALUES BACK TO ORIGINAL VALUES.
C
          DO 140 K = 1,NELMNT
          L = KIND(K)
          IF (L-4) 138,136,138
  136  ELVAL(K) = 1./ELVAL(K)
  138  CONTINUE
  140  CONTINUE
  200  CONTINUE
          RETURN
          END
```

Programs for Matrix Exponential Calculations (Chapter 8)

```
C
C PROGRAMS FOR EVALUATION OF THE MATRIX EXPONENTIAL
C
      SUBROUTINE MATEX   (A,EXPAT,N,T,NPOW)
C
C MATEX   COMPUTES THE MATRIX EXPONENTIAL, EXP(AT)
C A TRUNCATED POWER SERIES IS USED TO EVALUATE EXP(AT)
C TERMS UP TO AT**NPOW ARE INCLUDED IN THE POWER SERIES
C A IS NXN
C
      DIMENSION A(10,10),EXPAT(10,10),PEX(11)
      CALL POLMEX (PEX,NPOW,T)
      CALL MATPS (A,PEX,EXPAT,N,NPOW)
      RETURN
      END
C
      SUBROUTINE MATPS (A,P,F,N,NP)
      DIMENSION A(10,10),F(10,10),TEMP(10,10),AL(10,10),P(11)
C
C MATPS EVALUATES A MATRIX POWER SERIES
C THE SERIES COEFFICIENTS ARE GIVEN IN P
C P(1) CONTAINS THE COEFFICIENT OF THE ZEROTH POWER
C F IS THE SUM OF THE POWER SERIES
C F=SUM(P(I)*(A**(I-1)))FOR I=0 TO I=NP
C NP IS THE HIGHEST POWER OF A IN THE SERIES
C A AND F ARE NXN (SQUARE) MATRICES
C
      DO 10 I=1,N
      DO 10 J=1,N
      IF (I-J) 20,30,20
   30 F(I,I)=P(1)
      GO TO 10
   20 F(I,J)=0.
   10 CONTINUE
      IF (NP) 100,100,40
   40 CONTINUE
      DO 50 I=1,N
```

```
      DO 50 J=1,N
      TEMP(I,J)=A(I,J)
   50 F(I,J)=F(I,J)+P(2)*A(I,J)
      IF (NP-1) 100,100,60
   60 CONTINUE
      NP1 = NP + 1
      DO 80 L = 3,NP1
      CALL MATMUL (TEMP,A,AL,N,N,N)
      DO 70 I=1,N
      DO 70 J=1,N
      F(I,J)=F(I,J)+P(L)*AL(I,J)
   70 TEMP (I,J)=AL(I,J)
   80 CONTINUE
  100 CONTINUE
      RETURN
      END
C
      SUBROUTINE MATMUL (A1,A2,A3,N1,N2,N3)
C
C MATMUL MULTIPLIES MATRICES A1 AND A2.
C A1 IS N1 BY N2.
C A2 IS N2 BY N3.
C A3=A1XA2 IS N1 BY N3.
C ALL MATRICES MUST BE DIMENSIONED (10,10) IN CALLING PROGRAM
C
      DIMENSION A1(10,10),A2(10,10),A3(10,10)
      DO 20 I=1,N1
      DO 20 K=1,N3
      A3(I,K)=0.
      DO 20 J=1,N2
   20 A3(I,K)=A3(I,K) +A1(I,J)*A2(J,K)
      RETURN
      END
C
      SUBROUTINE POLMEX (PEX,N,T)
C
C POLMEX CALCULATES THE COEFFICIENTS IN A TRUNCATED SERIES
C FOR THE MATRIX EXPONENTIAL, EXP(AT)
C N IS THE HIGHEST POWER IN THE SERIES
C COEFFICIENTS ARE THE COEFFICIENTS OF THE POWERS OF A AND INCLUDE T
C
      DIMENSION PEX(11)
      N1=N+1
      FACTRL=1.
      PEX(1)=1.
      DO 10 I=2,N1
      PEX(I)=(PEX(I-1)*T)/FACTRL
      XI=I
   10 FACTRL=FACTRL*XI
      RETURN
      END
```

Routines Useful in System Analysis (Chapter 9)

```
      SUBROUTINE TRASTA
C TRASTA GENERATES A TRANSFER FUNCTION FROM A STATE MODEL
C
C TRASTA USES THE LEVERRIER ALGORITHM
C SEE MORGAN: SENSITIVITY ANALYSIS AND SYNTHESIS
C IEEE TRANS ON AUTOMATIC CONTROL, JULY, 1966
C
C************************************************************
C
C THE STATE EQUATION IS:
C
C            XDOT=A*X+B*U
C
C        X=STATE VECTOR (NX1)
C        U=CONTROL VECTOR (MX1)
C
C              Y=C*X+D*U
C
C        Y=OUTPUT VECTOR (LX1)
C
C************************************************************
C
      DIMENSION R(20,20), TEMP(20,20), TEMPV(21)
      DIMENSION PD(21), PN(21)
      COMPLEX POLES(20), ZEROES(20)
      COMMON/TRAN/PD,PN,ND,NN,POLES,ZEROES
      DIMENSION A(20,20),B(20),C(20),X(20)
      COMMON/SYST/A,B,C,D,N,X
C
C MAXIMUM CAPABILITY OF PROGRAM AS DIMENSIONED ABOVE IS 20TH ORDER SYST
C SINGLE INPUT-SINGLE OUTPUT
C MULTIPLE INPUTS OR OUTPUTS WOULD REQUIRE ITERATION OVER M OR L
C THIS ITERATION IS NOT PROVIDED IN THIS PROGRAM
C
      PD(N+1)=1.
      DO 3  J=1,N
      DO 3  K=1,N
```

```
    3  TEMP(J,K)=A(J,K)
       DO 100 I=1,N
C
C FORM THE TRACE OF TEMP AND STORE IN TR.
C
       TR = 0.
       DO 5 J = 1,N
       TR = TR + TEMP(J,J)
     5 CONTINUE
C
       XI = I
       PD(N+1-I)=-TR/XI
       DO 10  J=1,N
       DO 10  K=1,N
    10 R(J,K)=TEMP(J,K)
       DO 20  J=1,N
    20 R(J,J)=R(J,J)+PD(N+1-I)
C
C FORM THE MATRIX PRODUCT OF A AND R
C
       DO 30 II=1,N
       DO 30 JJ = 1,N
       TEMP(II,JJ) = 0.
       DO 30 KK = 1,N
       TEMP(II,JJ) = A(II,KK)*R(KK,JJ) + TEMP(II,JJ)
    30 CONTINUE
C
C MULTIPLY C BY R AND STORE TEMPORARILY IN TEMPV
C
       DO 40 J=1,N
       TEMPV(J)=0.
       DO 40 K=1,N
    40 TEMPV(J)=TEMPV(J)+C(K)*R(K,J)
C
C MULTIPLY C*R TIMES B AND STORE IN NUMERATOR POLYNOMIAL
C
       IF (N-I) 100,100,80
    80 CONTINUE
       PN(N-I)=0.
       DO 50 J=1,N
    50 PN(N-I)=PN(N-I)+TEMPV(J)*B(J)
   100 CONTINUE
C
C COMPUTE THE DOT PRODUCT OF B AND C AND STORE IN PN(N)
C
       PN(N) =0.
       DO 120 II = 1,N
       PN(N) = PN(N) + B(II)*C(II)
   120 CONTINUE
C
       PN(N+1) = 0.
       DO 60 J=1,N
    60 PN(N)=PN(N)+C(J)*B(J)
       ND = N + 1
       DO 70 J = 1,ND
       PN(J) = PN(J) + D*PD(J)
    70 CONTINUE
       NN = N
       RETURN
       END
```

STATRA is a routine to find a state representation from a transfer function representation, as discussed in Section 9.3.

```
    SUBROUTINE STATRA
```

```
C
C STATRA GENERATES A STATE MODEL FROM A TRANSFER FUNCTION
C THE STATE MODEL IS IN PHASE VARIABLE FORM
C FOR THE TRANSFER FUNCTION-
C PN IS THE NUMERATOR POLYNOMIAL OF DEGREE NN-1
C PD IS THE DENOMINATOR POLYNOMIAL OF DEGREE ND-1
C
C       Last Modified 8/14/85
C
        DIMENSION PD(21), PN(21)
        COMPLEX POLES(20), ZEROES(20)
        COMMON/TRAN/PD,PN,ND,NN,POLES,ZEROES
        DIMENSION A(20,20),B(20),C(20),X(20)
        COMMON/SYST/A,B,C,D,N,X
        DIMENSION PNTEMP(21)
C
C CHECK FOR EQUAL ORDER NUMERATOR AND DENOMINATOR.
C
        IF (NN .GT. ND) THEN
            WRITE (6,*)' No State Model generated.'
            WRITE (6,*)' Numerator Order > Denominator Order.'
            RETURN
        ENDIF
C
C Set top N-1 rows of A and set D to zero.
C D will be changed if NN=ND.
C
        N=ND-1
        N1=ND-1
        D = 0.
        DO 30 I=1,N1
        DO 30 J=1,N
        IF (J .EQ. (I+1)) THEN
            A(I,J) = 1.
        ELSE
            A(I,J) = 0.
        ENDIF
   30 CONTINUE
C
C NORMALIZE TRANSFER FUNCTION
C SET COEFFICIENT OF HIGHEST POWER IN DENOMINATOR EQUAL TO ONE
C
        DO 40 I=1,NN
   40 PN(I)=PN(I)/PD(ND)
        DO 50 I = 1,ND
   50 PD(I)=PD(I)/PD(ND)
        DO 60 I=1,N
   60 A(N,I)=-PD(I)
        IF (NN .EQ. ND) THEN
            D = PN(NN)
            NN = NN-1
            DO 65 I = 1,NN
                PNTEMP(I) = PN(I) - PN(NN)*PD(I)
   65       CONTINUE
            NN = NN + 1
        ENDIF
   66 CONTINUE
        DO 69 I=1,N
            X(I) = 0.
            C(I) = 0.
   69 CONTINUE
C
        DO 70 I=1,NN
            IF (NN .EQ. ND) C(I) = PNTEMP(I)
            IF (NN .LT. ND) C(I) = PN(I)
```

```
   70 CONTINUE
      IF (NN .LT. 20) THEN
          NN1=NN+1
          DO 90 I=NN1,20
   90     C(I)=0.
      ENDIF
  100 CONTINUE
      DO 110 I=1,N1
  110 B(I)=0.
      B(N)=1.
      RETURN
      END
C
C**********************************************************
C
```

Next we examine two routines, GAINMARGIN and PHASEMARGIN, both of which use the secant method to calculate gain or phase margin. Both call TRANEVAL to evaluate the transfer function at any given frequency. Data is passed to TRANEVAL through a COMMON block labeled "TRAN."

```
C
C**********************************************************
C
      SUBROUTINE PHASEMARGIN
      DIMENSION PD(21), PN(21)
      COMPLEX POLES(20), ZEROES(20)
      COMMON/TRAN/PD,PN,ND,NN,POLES,ZEROES
          WRITE (6,*)' Input the frequency range for the'
          WRITE (6,*)' Phase Margin search.'
          WRITE (6,*)' Input smaller frequency, then larger.'
          READ (5,*) FREQL, FREQR
          DO 240 I = 1,20
          CALL TRANEVAL (FREQL,G,PHI,GR,GI)
          FLEFT = 20.*(ALOG10(G))
          CALL TRANEVAL (FREQR,G,PHI,GR,GI)
          FRIGHT = 20.*(ALOG10(G))
          FREQZERO = FREQL-FLEFT*(FREQ-FREQL)/(FRIGHT-FLEFT)
          CALL TRANEVAL (FREQZERO,G,PHI,GR,GI)
          FZERO = 20.*(ALOG10(G))
             IF (FZERO .LT. 0.) FREQR = FREQZERO
             IF (FZERO .GT. 0.) FREQL = FREQZERO
             IF (FZERO .EQ. 0.) GO TO 250
  240     CONTINUE
  250 CONTINUE
          PHIM = PHI + 180.
          WRITE (6,*)' Phase Margin = ',PHIM,' at ',FREQZERO,' Hertz.'
      RETURN
      END
C
C**********************************************************
C
      SUBROUTINE GAINMARGIN
      DIMENSION PD(21), PN(21)
      COMPLEX POLES(20), ZEROES(20)
      COMMON/TRAN/PD,PN,ND,NN,POLES,ZEROES
          WRITE (6,*)' Input the frequency range for the'
          WRITE (6,*)' Gain Margin search.'
          WRITE (6,*)' Input smaller frequency, then larger.'
          READ (5,*) FREQL, FREQR
          DO 240 I = 1,20
          CALL TRANEVAL (FREQL,G,PHI,GR,GI)
          FLEFT = PHI + 180.
          CALL TRANEVAL (FREQR,G,PHI,GR,GI)
          FRIGHT = PHI + 180.
```

```
          FREQZERO = FREQL-FLEFT*(FREQR-FREQL)/(FRIGHT-FLEFT)
          CALL TRANEVAL (FREQZERO,G,PHI,GR,GI)
          FZERO = PHI + 180.
             IF (FZERO .LT. 0.) FREQR = FREQZERO
             IF (FZERO .GT. 0.) FREQL = FREQZERO
             IF (FZERO .EQ. 0.) GO TO 250
  240     CONTINUE
  250 CONTINUE
          GM = -20.*(ALOG10(G))
          WRITE (6,*)' Gain Margin = ',GM,' at ',FREQZERO,' Hertz.'
      RETURN
      END
C
C*****************************************************************
C
      SUBROUTINE TRANEVAL (FREQ,GMAG,PHI,GR,GI)
      DIMENSION PD(21), PN(21)
      COMPLEX POLES(20), ZEROES(20)
      COMMON/TRAN/PD,PN,ND,NN,POLES,ZEROES
      COMPLEX S,GAIN,RN,RD
C
C EVALUATE NUMERATOR AND DENOMINATOR AT S = J*OMEGA
C
      S=(0.,1.)*2.*3.1415927*FREQ
      CALL POLEVL (PN,NN,S,RN)
      CALL POLEVL (PD,ND,S,RD)
      GAIN = RN/RD
      GMAG=CABS(GAIN)
      GI = AIMAG(GAIN)
      GR = REAL(GAIN)
      PHI=ATAN2(GI,GR)
C
C CONVERT PHASE FROM RADIANS TO DEGREES
C
      PHI = 57.29578*PHI
      IF (PHI .GT. 0.) PHI = PHI - 360.
      RETURN
      END
C
C*****************************************************************
C
C
C*****************************************************************
C
      SUBROUTINE MATPOL (P,NP,A,N)
C
C MATPOL CALCULATES THE COEFFICIENTS OF THE CHARACTERISTIC EQUATION OF A
C A IS AN NXN MATRIX
C POLMAT USES BOCHERS FORMULA
C
      DIMENSION A(10,10), P(11), TEMP(10,10), AK(10,10), T(10)
      NP=N+1
C
C LOAD IDENTITY MATRIX IN TEMP
C
      DO 10 I=1,N
      DO 10 J=1,N
      IF (I .EQ. J) THEN
         TEMP (I,J) = 1.
      ELSE
         TEMP (I,J) = 0.
      ENDIF
   10 CONTINUE
C
C COMPUTE POWERS AND TRACES OF POWERS OF A
C
```

```
      DO 20 I=1,N
      CALL MATMUL (A,TEMP,AK,N,N,N)
      CALL MATTRC (AK,N,TRACE)
      T(I)=TRACE
      DO 25 J=1,N
      DO 25 K=1,N
   25 TEMP(J,K)=AK(J,K)
   20 P(NP-I)=0.
      P(NP)=1.
      DO 40 I=1,N
      INDEX = NP-I
   40 P(INDEX)=-T(I)
      N1=N-1
      IF(N1) 80,80,50
   50 CONTINUE
      DO 70 I=1,N1
      DO 60 J=1,I
      K=N-I+J
   60 P(N-I)=P(N-I)-P(K)*T(J)
      XI=I+1
   70 P(N-I)=P(N-I)/XI
   80 CONTINUE
      RETURN
      END
C
C************************************************************
C
      SUBROUTINE MATTRC (A,N,T)
      DIMENSION A(10,10)
C MATTRC COMPUTES THE TRACE OF THE SQUARE (N BY N) MATRIX,A.
C THE RESULT IS RETURNED IN T.
      T=0.
      DO 10 I=1,N
   10 T=T+A(I,I)
      RETURN
      END
```

Index

AC analysis, 18–19
ACSL, 73, 111
Algorithms:
  backward Euler, integration, 84–89, 106–108
  Bernoulli method, factoring, 244–251
  binary search, 33
  contraction mapping, 36
  derivative computation, 39
  directional search, optimization, 163–164
  Euler, integration, 66–67
  Fletcher-Powell, optimization, 165–167
  Gaussian Reduction, 17–18
  Gear, integration, 110–112
  Golden Section, optimization, 154–157
  gradient, optimization, 125–126
  Leverrier, 239–240
  linear system response:
    arbitrary inputs, 230–231
    step, ramp and sinusoid, 229–230
  matrix exponential, 220
  modified trapezoidal, integration, 98, 109–110
  Newton-Raphson, 39
  pattern search, optimization, 171–173
  predictor-corrector, integration, 103–105
    Adams-Moulton, 105
  quadratic jump, optimization, 160
  Runge-Kutta, integration:
    second-order, 96, 98
    fourth-order, 99–101
    Nth-order, 101
  simplex, optimization, 168–171
  trapezoidal, modified, integration, 108–110
  tree finding, 192–193
Autonomous linear systems, 219–225

Backward Euler integration algorithm, 84–89, 106–108
  companion models for circuits, 86–89
Bernoulli method, factoring, 245–250
Binary search, 33
Bocher's method, 228–229, 238
Branch, 184

Chaos, strange attractor, 57–58
Computation time, *see* Time, computation
Conductance matrix, 7
Contraction mapping, 34–36, 44, 105
Controlled sources, 23–28, 205–209
Convergence rate, 44
  Bernoulli method, 250
  gradient algorithm, 132–133, 137–138
  matrix exponential series, 218
  Newton-Raphson, 38
    diode circuit, 45–47
  secant and Newton-Raphson, 55

Derivative, computation of, 39–40

Equations:
  linear, simultaneous, 3
    numerical solution, 10
Error:
  derivative computation, 39
  Euler integration, 69–70
  multistep Euler integration, 82–84
Error propagation-simulation, 82–84
Euler integration algorithm, 62
  error, 69–70
  matrix exponential, 217